AF880547

DEFENCE RELATED SME'S

NATO Science Series

A series presenting the results of scientific meetings supported under the NATO Science Programme.

The series is published by IOS Press and Kluwer Academic Publishers in conjunction with the NATO Scientific Affairs Division.

Sub-Series

I.	Life and Behavioural Sciences	IOS Press
II.	Mathematics, Physics and Chemistry	Kluwer Academic Publishers
III.	Computer and Systems Sciences	IOS Press
IV.	Earth and Environmental Sciences	Kluwer Academic Publishers
V.	Science and Technology Policy	IOS Press

The NATO Science Series continues the series of books published formerly as the NATO ASI Series.

The NATO Science Programme offers support for collaboration in civil science between scientists of countries of the Euro-Atlantic Partnership Council. The types of scientific meeting generally supported are "Advanced Study Institutes" and "Advanced Research Workshops", although other types of meeting are supported from time to time. The NATO Science Series collects together the results of these meetings. The meetings are co-organized by scientists from NATO countries and scientists from NATO's Partner countries – countries of the CIS and Central and Eastern Europe.

Advanced Study Institutes are high-level tutorial courses offering in-depth study of latest advances in a field.
Advanced Research Workshops are expert meetings aimed at critical assessment of a field, and identification of directions for future action.

As a consequence of the restructuring of the NATO Science Programme in 1999, the NATO Science Series has been re-organized and there are currently five sub-series as noted above. Please consult the following web sites for information on previous volumes published in the series, as well as details of earlier sub-series:

http://www.nato.int/science
http://www.wkap.nl
http://www.iospress.nl
http://www.wtv-books.de/nato_pco.htm

Series V: Science and Technology Policy - Vol. 43 ISSN 1387-6708

Defence Related SME's

Analysis and Description of Current Conditions

Edited by

Fernando Duarte Carvalho

INETI, Lisbon, Portugal

IOS
Press

Amsterdam • Berlin • Oxford • Tokyo • Washington, DC

Published in cooperation with NATO Scientific Affairs Division

Proceedings of the NATO Advanced Research Workshop on
Defence Related SME's – Analysis and Description of Current Conditions
23–26 June 2002
Funchal, Portugal

© 2004, IOS Press

All rights reserved. No part of this book may be reproduced, stored in a retrieval system, or transmitted, in any form or by any means, without prior written permission from the publisher.

ISBN 1 58603 408 1
Library of Congress Control Number: 2004100157

Publisher
IOS Press
Nieuwe Hemweg 6B
1013 BG Amsterdam
Netherlands
fax: +31 20 620 3419
e-mail: order@iospress.nl

Distributor in the UK and Ireland
IOS Press/Lavis Marketing
73 Lime Walk
Headington
Oxford OX3 7AD
England
fax: +44 1865 75 0079

Distributor in the USA and Canada
IOS Press, Inc.
5795-G Burke Centre Parkway
Burke, VA 22015
USA
fax: +1 703 323 3668
e-mail: iosbooks@iospress.com

LEGAL NOTICE
The publisher is not responsible for the use which might be made of the following information.

PRINTED IN THE NETHERLANDS

Preface

Today, an area where change is particularly apparent is that of the relationships in and between countries in matters of national defence. Gone are the huge engines of military industrial might, particularly in the former Soviet Union, and to some extent in Western Europe, and struggling in their wake are networks upon networks of small and medium-sized enterprises (SME's), groping with the harsh reality of survival in this new world. For the first time, many if not most of them, are being forced to deal with what are to them new issues such as new product development, marketing for the civilian world, private-sector financing and adopting competitive business practices and generally living outside the shadow of the defence business. Crucial at this juncture is to provide forms of assistance to guide these SME's over new ground. Most observers agree that most businesses are small or medium in size, in the sense that they are not large corporations. These businesses are the lynchpin of stable economic conditions and ultimately provide essential products and services to users. The failure of SME's to convert successfully to civilian markets would cause a devastating blow to national economies. Accordingly, the need and importance of dealing with these defence conversion issues cannot be overstated.

This book presents a collection of papers on the topic of defence-related small and medium enterprises and the challenge of readjustment presented to them in the 21st century. The papers were first given at a NATO Advanced Research Workshop held in Madeira in June 2002, organised by General Pelágio Castelo Branco, Professor Naum Yakimoff and Mr. Thomas Shea.

The organisers gratefully acknowledge the financial support of the Scientific and Environmental Affairs Divison of NATO in the preparation of the workshop. Thanks are also expressed to the NATO Science Panel Observer, Professor Luigi Tomasini, to the Regional Government of Madeira and to the DANOTEC Association for support during the workshop.

Finally, the organisers express their gratitude to all the participants, who were the principal actors.

The Editor wishes to thanks Liz Cowan for her work in reviewing this book.

FDC
June 2003

Introduction

Dynamic and activist… those are the words which best characterize the Advanced Research Workshop (ARW) held in Portugal from 23 to 26 June 2002. For four intense days, 31 participants from 18 nations (including a Delegation of 3 high-ranking Russian Officials, although lamentably missing the representatives from Lithuania and Uzbekistan due to air strikes), talked about, described and discussed small and medium enterprises (SME's); how they can adapt to change, and how they play such a crucial role in business, indeed global, development. A common theme running through our discussions was that SME's, even in diverse nations, have more similarities than differences. Accordingly, speakers could relate well to each other, and those with development interests in south-eastern Europe profited greatly from the proceedings.

The highest priority for this workshop was to provide maximum opportunity for participation by all persons, through the following format: Sessions ran from one to two hours, with three or four speakers, each of whom spoke for 15-25 minutes. Visual aids were either overhead slides or PowerPoint presentations, followed by a question and answer period. There were normally two or three sessions in the morning, and two or three in the afternoon. Any observer would be impressed with the easy manner in which participants could talk with and question (even argue with) each other, always in a constructive way.

The highlight of the workshop was the last day's session when all participants got a chance to answer the following question: what subject would you like to have discussed at this workshop? Nobody was bashful, and the moderator listed on a huge paper easel at least two dozen different topics. There was a widespread feeling of satisfaction that each person's opinion mattered. Next, participants were asked to state what barriers existed to achieve stated goals or topics. These proceedings were verbal, and the moderator translated them into succinct bullets, which were written down on a separate easel. Finally, each participant was told to write what he or she wanted to see come out of this Workshop in terms of future NATO-supported actions. Although there were 28 separate suggestions, many were similar, some very much so. The moment of truth came next when each participant was instructed to select one, not two…but one only, for a future NATO-supported action or event. The near-unanimous Workshop recommendations are set out below.

As a closing remark, it is worth mentioning that the attending NATO Advisory Panel observer and a few ARW veteran participants all expressed admiration for and approval of the dynamics shown at this NATO-Russia Cooperation event and the easy way in which ideas were shared and analyzed.

Conclusions

The resulting Recommendations to NATO, supported by a number of Expectations, Challenges/Barriers and Actions, stem from the wide-ranging scope of the discussions held by the 31 Participants in the 7 Sessions of the Workshop.

Of special note, however, is the outstanding parallel contribution made by the Russian Official Delegation through the following ideas and assertions, which are likely to offer much food for thought: the defence industry in the Russian Federation in general and the SME's utilizing dual use technologies in particular, are living through a period of vast reforms, the objective of which is to transform these SME's into a main source of technological innovation in the science-oriented sector of the economy. The Russian Federation is interested in deepening cooperation with NATO and EU Member Countries in the field of high technologies.

The Workshop held in Portugal turned out to be a very useful and timely forum of ideas and information in this regard. We really hope that the recommendations worked out by the participants will help us to mutually foster a profitable cooperation between SME's and prime companies of our countries, with expert participation and guidance of NATO.

Recommendations

1. Training programme (NATO Advanced Training Course).
2. Internship programme East-West (NATO Internship in East-West Science Policy, PHARE, ISP.
3. Definition, issues, aspects, etc., related to SME's in defence, through further involvement of NIAG and other organisations, including funding. (NSIP, PfP, NIAG, C4, Industry, Primes).

Expectations

- Bridging Government - SME
- Public-private partnership – SME's/Primes
- Bridging Government - Primes
- SME success in defence
- More efficient knowledge/TT
- NATO focus on SME's
- Civilian - defence SME's
- Partnership
- Business opportunities for SME's in Central/Eastern Europe
- Processes for developing SME - NATO role
- SME's in SMC's and SME's in countries with primes
- Better understanding of offsets
- Virtual company "do as we do"
- Programmes - rules
- NIAG as a forum
- Learning experience from other countries
- How can Slovakian SME's (as an example) be involved in the defence industry
- IT for SME's development
- Integration of CEE countries in Euro-Atlantic systems
- Actions, programmes, proposals
- Civil primes – SME's

Challenges/Barriers

- Government – SME's
- Prime – SME's
- Synergy - Government/Academia/SME's/Industry
- Lack of transparency and access to information

- Lack of collaboration between Government-NGO-Social partner
- Bureaucracy
- Difference in priorities Government/SME's/Prime
- Access to capital
- No formal planning
- Cultural and organisational differences
- Lack of confidence of SME's in getting into partnership
- Human resources in SME development
- Lack of life-long learning for entrepreneurs
- Harmonisation of IP law
- Cost of IP protection for SME's
- Too much uninformed government interaction
- Lack of consultants in SME development
- Export regulations
- Lack of knowledge on SME support programmes and NATO programmes
- Protection of national businesses
- Influence of the primes on decision making/shareholders
- How to develop a closer relationship with Russians
- Gap between CEE and developed countries
- Gap between secrecy and standards
- Unfair competition between US and EU primes, subs
- Management of expectations
- Lack of definition and policy of dual use
- Common SME solutions
- Psychology/mentality
- Lack of SME participation in supply chains
- Time and resources of SME's involved in (NATO) programmes
- Corruption

Actions

- Use of offsets (and countertrade) to implement long term relations between East-West SME's
- NATO alignment and definition of national policies for SME's
- One-stop virtual shop (NATO) information and assistance for SME's
- International training programmes for SME managers/trainers/community leaders/government officials/academia/specific projects/partnering programmes (East-West)
- Common C4 architecture development and ATD
- Support SME's in sustainable development
- Increasing awareness of SME's about NATO programmes in Partner countries
- Assisting retired military personnel to be active in SME's (start-ups)
- NATO preferences in procurement for CEE SME's and clusters
- Further involvement of NIAG and more money in definition/issues/aspects
- Cooperation between SME's in dual use
- Study on SME's
- Setting up SME's
- Database of SME's with developing technologies
- Internship programmes East-West

Mr. **Richard A. BENDIS**,
Innovation Philadelphia,
2600 Centre Square West,
1500 Market Street,
Philadelphia, PA 19102, USA

Prof. **Fernando D. CARVALHO**,
INETI,
Rua Brig. Fernando Alberto Oliveira,
Bloco 2, 3A,
2615-007 Alverca, Portugal

General **Pelágio CASTELO BRANCO**,
DANOTEC,
Av. de Moscavide 18,
1885-998 Moscavide, Portugal *
(*new address:
C3P Centre for Pollution Prevention,
Rua da Conçeicão, 45 – 1E,
1100-151 Lisbon, Portugal)

Prof. **Kresimir COSIC**,
Croatia Centre for Strategic Studies and
Faculty of Electrical Engineering,
University of Zagreb, Unska 3,
10000 Zagreb, Croatia

Mr. **Heinz FIEDLER**,
The SPICE Group,
Zoobogen,
Hardenbergplatz 2,
10623 Berlin, Germany

Dr. **Gennadi GAIKO**,
BSBI Association,
5 Minskaya Street,
Kolodishchi,
223051 Minsk, Belarus

Dr. **Natalia KOVALEVA**,
Centre for Science Research and Statistics,
11 Tverskaya Street,
103905 Moscow, Russia

Mr. **Andrey A. KRUKTO**,
Chief of NATO Division,
Ministry of Foreign Affairs,
Moscow, Russia

Prof. **Ian McFARLANE**,
Cranfield University, UK
(The Lowlands, Division Lane,
Blackpool, FY4 5EA, UK)

Dr. **Péter MOGYOROSI**,
Laser Consult, Inc., Pf. 1191,
6701 Szeged, Hungary

Mr. **Brian MORRIS**,
The European Association of
Aerospace Manufacturers (AECMA),
Gulledelle 94,
1200 Brussels, Belgium

Mr. **Guy SERVOLLE**,
Dassault Aviation,
Rue de Strasbourg 13,
1030 Brussels, Belgium

Dr. **Vladimir SHABAROV**,
Defence Industries Department,
Ministry of Science and Technology,
Miussjaya Sq, 3,
125319 Moscow, Russia

Dr. **Velizar SHALAMANOV**,
George C. Marshall Association,
3 Sheinovo Street,
1504 Sofia, Bulgaria

Mr. **Thomas F. SHEA**,
Office of Economic Adjustment,
Department of Defense,
400 Army Navy Drive,
Arlington, VA 22202, USA

Dr. **Mircea SLANINA**,
ICPE – Trafil AS,
Bd. Mangeron 49,
6600 Iasi, Romania

Eng. **João P. TABORDA**,
INTELI, Estr. Paço do Lumiar,
Campus do INETI, Edif. 198,
1649-038 Lisbon, Portugal

Dr. **Todor TAGAREV**,
Centre for National Security
and Defence Research,
Bulgarian Academy of Sciences, 1,
15 November Street,
1040 Sofia, Bulgaria

No photo available

Prof. **Trevor TAYLOR**,
Royal Military College of Science,
Cranfield University,
Shrivenham,
Swindon, SN6 8LA, UK

Prof. **Luigi TOMASINI**,
Dip. di Economia Politica,
Universita di Siena,
Piazza S. Francesco 7,
53100 Siena, Italy

Mr. **Rene TÕNNISSON**,
Tartu Science Park,
Riia 185,
51014 Tartu, Estonia

Dr. **Stefan VRATNY**,
BIC Bratislava, Zochova 5,
81103 Bratislava,
Slovak Republic

Mr. **Han WILLEMS**,
Netherlands Defence
Manufacturers Association (NIID),
The Hague, The Netherlands

Prof. **Naum YAKIMOFF**,
Scientific Secretary General,
Bulgarian Academy of Sciences,
1, 15 November Street,
1040 Sofia, Bulgaria

Dr. **Faruk A. YARMAN**,
Department of Computer Engineering,
Baskent University,
Ankara, Turkey

Prof. **William ZADORSKY**,
Department of Chemical Engineering,
Ukrainian State University, Suite 23,
Building 41/43 Komsomolskaya Street,
Dnepropetr 2, 320002 Ukraine

Dr. **Kryzstof ZASIADLY**,
Polish Business and Innovation
Centres Association,
Rubiez 46,
61-612 Poznan, Poland

Dr. **Aleksander M. ZHEREBIN**,
State Centre of Systems Research,
7 Victorenko Street,
125319 Moscow, Russia

Invited but unable to attend:

Dr. **Pranas B. MILIUS**, Innovation Centre, Kaunas University of Technology, K.Donelaico 73, 3006 Kaunas, Lithuania

Prof. **Gulnara RASHIDOVA**, Technology Transfer Agency, 29 Soleimanova Street, 700017 Tashkent, Uzbekistan

Present at the opening and closing sessions:

Dr. **Carlos PEREIRA**, Vice-President, Regional Government of Madeira, Av. Zarco, 9004-527 Funchal, Madeira

Dr. **Nuno TEIXEIRA**, Vice-President, Regional Government of Madeira, Av. Zarco, 9004-527 Funchal, Madeira

Photos of the Session

Photo 1 – Chairmen's Table: Mr. Thomas F. SHEA, Gen. Pelágio CASTELO BRANCO, Prof. Luigi TOMASINI and Prof. Naum YAKIMOFF

Contents

Defense Related SME's: Analysis and Description of Current Conditions
F.D. Carvalho (Ed.)
IOS Press, 2004

Technology-based Economic Development

Richard A. BENDIS
President and CEO
Innovation Philadelphia, USA

Abstract. Economic evolution of different countries and regions occurs at various stages. But there is an element within each of these varying locales that has the potential to unify us all: knowledge. Knowledge can be the nexus for all of us to create a knowledge-based society. Economic evolution started with the Agriculture Age and progressed to the Industrial revolution. We now find ourselves in the information age, desiring to transform information into a knowledge-based society. Given that the 21st Century World has changed dramatically and is much more complex due to a litany of factors:

- distributed intelligence
- complex technologies
- diverse workforce
- creative transformation
- continuous innovation
- global economy
- infrastructure renewal
- sustainable development
- the need for career-long learning
- finite resources

We must embrace the fact that science is the fuel for technology's engine and that technology is the engine of economic growth. Because in the end, technology is essential to the creation of jobs, to the building of new industries, and to the improvement of our standard of living.

1. Global Recipe for Success

In order for a country to be competitive in the global economy, there are several worldwide competitiveness factors that must be taken into account. A country's infrastructure must be firmly grounded and well established to the extent to which its resources and systems are adequate to serve the basic needs of business. The scientific and technological capacity of a given country must be capable to successfully commercialize basic and applied research. Furthermore, these enterprises must be managed in an innovative, profitable, and responsible manner in order to ensure competitiveness.

The availability and qualifications of human resources are also key ingredients needed to prepare this global recipe for success. Once this macroeconomic understanding and evaluation of the domestic economy and what it takes to make it strong has been completed, the extent to which government policies are conducive to competitiveness must be evaluated within a national and international context. Within this global economy, the extent to which a country participates in international trade and investment flows plays a significant role in that country's prosperity, specifically the performance of the capital markets and quality of financial services that fuel that country's economy.

Human connectivity and communications networks have the ability to transform economic, political, and social relationships on a global scale. In the past, organizations strategized to gain competitive advantage. But the emphasis in the future will be to gain cooperative advantage. A core competency needed in individuals and organizations alike is connectivity with one another, namely in the form of public/private partnerships.

Progress is promoted by strong industry, government and university leadership, and sustained by dynamic public/private partnerships.

These new public/private leaders create new, responsive models of governance. To this end, the missions of education, industry and government must be seen as being inseparable. While education focuses on research, service, and lifelong learning; and industry focuses on product development, business processes, and generating a profit; and government focuses on economic benefits, returns on public investment, and sustainable development; we must find overlapping goals between all these ostensibly distinct partners in order to build successful public/private partnerships.

The government can play many roles in using science and technology to stimulate the economy including: long- term vision and planning; the identification of gaps and trends in the science and technology environment and the development of programs to fill these gaps; becoming a catalyst through strategic investments and public/private partnerships; the encouragement of a balanced and flexible research and development portfolio; the encouragement of private sector innovation, and the establishment of performance-based research and development. The specific roles of the national or federal government include:

- the maintenance of a healthy educated public
- job creation, economic health and security
- leadership in science, engineering, mathematics, and technology
- improved environmental quality and sustainable development, harnessed information technology; and enhanced national security.

The role of Industry is simple, "Wealth Creation!" "Capitalism is a process of Creative Transformation. The interaction of technological innovation with the competitive marketplace is the fundamental driving force in capitalist industrial progress", as stated by Joseph A. Schumpetter in 1942.

The roles of Academe are many, be they oriented with education and research, knowledge integration, resource investment, knowledge creation, continuous learning and innovation, knowledge transfer. There are many challenges for 21st Century Academe which has to educate students to see the world whole; sense the coupling among seemingly disparate fields of endeavor; perform synthesis in balance with analysis; build connections between the world of learning and the world beyond; and most importantly INNOVATION.

In order to strengthen the partnership of University-Industry-Government, there must be greater integration of efforts, increased resource sharing, expanded information flow, and more people exchange, to share knowledge and experience.

2. The Keystone State

With a land area of 44,817 sq. miles, the state of Pennsylvania, aside from being one of the original 13 colonies, has a rich and diverse history. From the city of Hershey, PA, which is known as the "Chocolate Capital of the World" to the first Daily newspaper that was published in Philadelphia in 1784, Pennsylvania has always played an important part in the education and cultural development of the United States of America. With 146 public colleges & universities and 589,838 full-time college students, it is no wonder that the following scientific and technological accomplishments occurred in this great state:

- First American hospital, Pennsylvania Hospital was founded in 1771
- First Piano in America was built under the name "Piano Forte" in 1775
- First Children's Hospital (CHOP) was founded in 1855
- First Medical School and teaching hospital, University of Pennsylvania was founded in 1874
- First Independent medical research facility in the U.S., the Wistar Institute was founded in 1892
- First Cancer Hospital, Fox Chase Cancer Center was founded in 1904
- First College of Pharmacy, University of the Sciences was founded in 1921
- First Computer was built in Philadelphia in 1946.

Located in the sixth largest state in the continental United States, from which 26% of the U.S. population is within one-day's drive and 50% of the U.S. population is within two hours by plane, the city of Philadelphia is situated in an ideal spot between the center of private capital of New York and the center of public capital of Washington, D.C, playing its role as the "buckle on the money belt." As for the city of Philadelphia itself, the facts and figures paint an equally as impressive picture when compared to the state. Located in the sixth largest metropolitan area (PA, NJ, DE) in the U.S. with roughly 6.2 million inhabitants and ranking as the fifth largest city in the U.S. with a population of 1.4 million, it is not surprising that the city has the sixth largest GMP in the U.S. with $182.4 million dollars.

Like the state, Philadelphia is also centered on education, given that it possesses the largest per capita concentration of higher education institutions in the U.S. with more than 85 colleges. As far as industry is concerned, eighty per cent of the world's major pharmaceutical companies (e.g. Merck, GlaxoSmithKline, Bristol Myers-Squibb, Johnson & Johnson, AstraZenaca, and Dupont) are within a two-hour drive of Philadelphia. Furthermore, with over 450 foreign-owned companies and subsidiaries employing over 60,000 in the region and with 34 foreign consuls and with its seven sister cities in the Sister International cultural exchange program, the Greater Philadelphia region maintains a very accessible and amicable relationship with the international community.

Selected as one of America's 32 most livable cities, Philadelphia is also home to the busiest port in the North Atlantic and the fastest-growing U.S. airport with 544 domestic flights and 42 international flights daily. Not to mention that the University City Science Center ranks as one of the top 10 research parks in the U.S. and the Metro Area appeared on the forth-annual world-class communities list. But that is not all. Besides its accessibility, Philadelphia is home to world-class art. The new $245 million dollar Kimmel Center houses the Philadelphia orchestra, dance, ballet, and theater, while the museums that inhabit the city range from the Philadelphia Museum of Art, the Barnes Foundation, the Rodin Museum, and the Franklin Institute.

3. What is Innovation Philadelphia?

Innovation Philadelphia (IP) is a quasi-private entity created to position Philadelphia as a leading global innovation region led by an initial Board of Directors that includes: Judith Rodin - Chairperson, President of the University of Pennsylvania; J.P. Garnier - Vice Chair, President & CEO of GlaxoSmithKline; Brian Roberts - Vice Chair, President of Comcast; Mayor John Street - Ex Officio, City of Philadelphia; and Richard A. Bendis, President & CEO of Innovation Philadelphia.

IP's mission is to enhance the Global Innovation Economy of Philadelphia through technological leadership.

With Philadelphia's Strategic Technology Clusters including:

- pharmaceutical
- bioscience/biotechnology
- healthcare
- information technology
- systems integration
- nanotechnology
- financial services
- Petrochemical
- communications and media

IP has developed eight objectives that will positively impact entrepreneurial development in the region.

GOAL: To increase the entrepreneurial economic development resources in the region through the development of the Innovation Philadelphia 2002 Entrepreneurs' Resource Guide.

GOAL: To increase the amount of pre-seed and seed capital investment in the region via the creation of the Mid-Atlantic Angel Fund.

The following outline describes the various stages of financing a company that would result.

1. *The Early Stage*

Initial (Seed)- A relatively small amount of capital provided to an investor or entrepreneur, usually to prove a concept. This may involve product development, but rarely involves initial marketing.
First (Start-up)- Financing provided to companies that have expended their initial capital and require funds, often to initiate commercial manufacturing and sales.

1. *Later Stage*

Second- Working capital for the initial expansion of a company that is producing and shipping and has growing accounts receivable and inventories. Although the company has clearly made progress, it may not yet be showing a profit.
Third- Funds provided for the major growth of a company whose sales volume is increasing and that is beginning to break even or turn profitable. These funds are typically for plant expansion, marketing, and working capital development of an initial investment.
Follow-on- a subsequent investment made by an investor who has made a previous investment in the company—generally a later stage investment in comparison to the initial investment.
Buyout- Funds provided to enable operating management to acquire a product line or business, which may be at any stage of development, from either a public or private company.

Bridge/Mezzanine- financing for a company expecting to go public usually within six months to a year. Often bridge financing is structured so that it can be repaid from proceeds of a public offering.
IPO- Initial Public Offering. A company's first offering of stock to the public.

GOAL: To increase the research investment in strategic technologies for the regional economy from federal, state, private, and industrial resources.

The accomplishment of these goals is actualized by the launch of the Innovation Philadelphia Federal Funding Program to Support Small Business Innovation Research (SBIR), Small Technical Transfer Research (STTR), and Advance Technology Program (ATP). The U.S. Small Business Innovation Research Program (SBIR) funds roughly $1.5 billion a year in early-stage R&D projects at small technology companies. SBIR funds technology development at an earlier stage than the private sector generally does, providing up to $850,000 in pre-prototype R&D to small technology companies. Most importantly, small companies retain the patent rights to any inventions they develop under these programs. The funding is awarded competitively, but the process is streamlined and user-friendly.

The purposes of the U.S. SBIR Program are: to stimulate technological innovation, to use small businesses to meet federal research and development needs, to increase private sector commercialization of innovations derived from federal research and development, and to foster and encourage minority and disadvantaged participation in technological innovation.

The United States Advanced Technology Program (ATP) attempts to bridge the gap between the research lab and the market. ATP has strict cost-sharing rules, and joint ventures must pay at least half of the project costs with a range of funding from $800,000 to $21 million per award in PA. ATP does not fund product development, and private industry bears the cost of product development, production, marketing, sales and distribution. Moreover, ATP awards are based on peer-review competitions with selection based on: innovation, technical risk, economic benefits to the U.S., and strength of the commercialization plan. All projects have goals, specific funding allocations, and completion dates established at the outset.

Innovation Philadelphia plans to support the ATP through conferences and information, technical assistance, commercial assistance, networking, market and technical research, draft proposal evaluation, and proposal preparation. IP could potentially support the SBIR and ATP grants with up to $4,000 in financial assistance for Phase I or II proposal preparation, with a bridge loan of up to $50,000 for bridging between Phase I & II, and with a CSBIR of up to a $300,000 matching investment for selected SBIR topics with commercial potential.

GOAL: To facilitate collaboration among Industry, Government, Academia, and Non-Profit Organizations involving key economic initiatives.

To accomplish this goal, we at IP are in the middle of developing the Greater Philadelphia Resource Guide. To combat the "brain drain" that has afflicted Philadelphia in the past, IP is developing a balanced portfolio of programs that supports Retention, Recruitment, and Grow Your Own strategies for the new economy in the region. To solve this problem, we are recruiting and retaining technology companies in the region.

The Pennsylvania Life Sciences Greenhouse is a $100 million dollar joint marketing and cooperative business attraction funded by the state. It has created a shared corporation in order to direct training, cooperative degree programs, and statewide outreach to K-12 system. It has also created a PA Life Science Passport to provide a single gateway to major research centers as well as a statewide Clinical Trials Network and a collaborative outreach to rural areas, minority businesses, and the manufacturing community. Its impact on the Philadelphia region will manifest itself by engendering 2,200 jobs directly and 6,500 indirectly. It will be governed by a 15-member board, with an investment of $33.8 million as its core grant with a $55 million match, and its focus will lie in the fields of biopharmaceuticals and bio-informatics.

The Port of Technology and the University City Science Center were established in 1963 and are currently owned by 31 academic and research institutions maintaining a mission to build and grow the knowledge-based economy by capitalizing on the intellectual resources of its stakeholders in order to create regional wealth. The value proposition of the Port and the UCSC are its core objectives: to accelerate, to increase chance of success, to conserve cash, and to eliminate distractions. At the Port of Technology, there exists a plug-and- play environment with two floors dedicated to technology startups, first-class wet labs, high-speed Internet connection, and a collaborative environment. (Email the Port at www.portoftech.com).

In order to reduce the "BRAIN DRAIN" and assist in the development of a world class life-long learning environment, Innovation Philadelphia has formed the Knowledge Industry Partnership (KIP), whose members include most of the local colleges and universities as well as numerous civic and government groups. To enhance the global image of Philadelphia as a leader in the "Knowledge Economy," IP has partnered with the United Nations, NATO, the City of Philadelphia's Chinese Trade Mission, and the Port of Technology on global initiatives. Our GLOBAL GATEWAY boasts International Alliances with:

• Argentina	• Japan
• Australia	• Korea
• Canada	• Spain
• Chile	• Russia
• China	• Mexico
• Czech Republic	• Ukraine
• Germany	• Thailand
• India	• Singapore
• Ireland	• Venezuela
• Israel	• United Kingdom

To accelerate the rate of commercialization of market-driven technologies and support university research and technology transfer, IP has formed a partnership with the Commercialization Working Group to assist the Port of Technology and the University City Science Center in the commercialization of strategic technologies to the region. To ease this transfer process, we promote the use of Innovation and Commercialization Corporations (ICCs) which provide management services directly to client companies including: assessment of technology for technical and economic viewpoints, analysis of market potential and barriers to entry, guidance in business plan development and strategic planning, supplementation to client's management resources until self-sufficient, determination of company's value, assistance in obtaining debt and equity financing, government grant assistance including SBIR/STTR/ATP/ARMF, and guidance in managing intellectual properties, including licensing.

With partners including Drexel, Innovation Philadelphia, Port of Technology, Temple University, Thomas Jefferson University, and the University of Pennsylvania, the

Commercialization Working Group (CWG) is a collaboration to accelerate the commercialization of new technology and the life sciences in the Greater Philadelphia region at the Port of Technology. The CWG is a partnership, which hopefully will leave the old paradigm behind and assume the new paradigm of the science and technology innovation process.

4. Come Invent The Future

Old Paradigm/Language	New Paradigm/language
Job creation	Wealth creation
Funding	Flexible investment
Individual	A balanced portfolio of investments
Funding research projects	Investing in commercialization of technology
Managing the budget	Maximizing the return on investment
	Entrepreneurial environment
Inputs and Activities	Results and outcomes
Linear commercialization process	Dynamic innovation process
Economic Development	Sustainable development
Technology Push	Market pull
Co-Funders	Stakeholders leveraging resources
Federal programs	National systems
Fragmented Activities	Integrate programs
Parochial state/industry interests	Clusters
Directing or steering	Creating partnerships
Regions being dedicated to regarding Federal/National programs	Regions involved in designing new science & technology programs

In order to implement this new paradigm effectively, you must possess willingness to deviate from traditional and parochial perspectives; a willingness to encourage public investment and risk taking; desire to develop trust through collaboration; ensure the paradigm is responsive to partners' missions; and want to build consensus of all constituents through education and participation.

The lessons that we have learned, i.e. what has worked in the past, are: that University/Industry/Government partnerships work; that enlightened self interest is more successful when it seeks mission overlap rather than mission change; that the matching of investments shares the costs and the benefits between all interested parties; the selection of technologies is strategic to partnering; the focus of investments on a few strategic technologies maximizes results; and the use of the sunset funding mentality ensures that graduates move out.

"Somebody has to do something, and it's just incredibly pathetic that it has to be us."
Jerry Garcia of The Grateful Dead

All of the puzzle pieces must come together early in order for the project to have any hope of commercial success.

- Market-driven opportunity
- Matching Research funding
- Scientific research team
- Federal grants
- Risk funding
- Private/Public/University Partnership
- Qualified management
- Commercialization partner
- Strategic implementation, market research, and marketing plan

Why is it that Innovation Philadelphia will work? Integrated science and technology collaboration for Philadelphia, IP has private sector leadership and essentially functions as a business. By successfully managing a technology investment portfolio for ROI, maintaining its operational flexibility, and living up to the standards of accountability with measurable outcomes, there is no doubt that Innovation Philadelphia will be able to reposition the Greater Delaware Valley region as a bastion of entrepreneurial success.

"It is not the strongest of the species that survive, nor the most intelligent, but the ones most responsive to change"
Charles Darwin.

F.D. Carvalho (Ed.)
IOS Press, 2004

Strategy of Small Defence Oriented Enterprises in a Time of Defence Budget Downsizing -Croatian Case Study-

Prof.dr.sc. Kresimir COSIC LTG(ret),
University of Zagreb
Faculty of Electrical Engineering and Computing, Unska 3,
10000 Zagreb, Croatia

Abstract. A relatively large defence industry during the Cold War period has been significantly reduced in many transitional states in the last decade due to heavy defence budget downsizing. These heavy budget cuts caused the collapse of many large defence corporations. Furthermore, since the beginning of the 21st century in the post-Cold War period many transitional states do not need large defence industrial facilities and large defence enterprises for self-protection. But protection of their small and medium size defence-oriented enterprises (SME's) is important for protection of their economies, employment, social stability etc. In such situations, transformation and adaptation of these SME's to the new business environment and conditions is imperative for their survival. Survival strategy of these small defence related companies requires: increase of their international competitiveness, acquiring of new dual use knowledge, expertise and technologies, stronger orientation to the civilian markets, identification of a new niche in civilian and defence markets, education and shaping of future customers, acceptance of a new management and organizational culture which supports innovation and creativity, minimization of bureaucracy, a modern type of personnel management and diversification of production. Competitiveness of these SME's on the globalized markets will depend, first of all, on their efforts to increase their R&D activities and to put more R&D based innovations into their new products. The Croatian case study given here is related to the business strategy formulation of one small R&D based Croatian defence-oriented enterprise which is trying to survive in a time of heavy defence budget cuts through transformation of its old mainly defence-oriented business strategy and products to a new one.

Introduction

In the post-Cold War period in Eastern and Central Europe, many huge military industrial state owned companies and facilities collapsed, while many of the SME's are still searching for business survival in a time of heavy defence budget downsizing. For the first time, most of these companies are being forced to deal with new issues and challenges, such as: new product development, marketing to civilian customers, private-sector financing, adopting new competitive business practices, living outside the shadow of the defence business, etc. The failure of these SME's to reach a successful adaptation of their services and products to the new defence and civilian markets would cause devastating consequences for their national economies and national security of these states. Therefore, the importance of dealing with transformation and adaptation processes and conversion issues of these SME's cannot be overstated. It is also very important to explore new ways in which these affected SME's can adjust their core businesses to the new economic situations, characterized by the heavy defence budget downsizing. In a case of successful transformation and adaptation, these SME's may play an important role in the economies of their countries and can make a substantial contribution to the creation of new jobs and prevention of social and political crises in these states.

On the basis of hypothetical statistics, we can guess that approximately up to 5% of these former defence oriented SME's have a chance to survive and to transform themselves into the new modern, internationally competent and competitive SME's, which will be capable of competing in the international defence industry business. Another 10 % of them have a chance to survive as in-state maintenance providers, while the rest will be definitely lost. This article concentrates mainly on the first group of only a small percentage of these SME's, which can transform themselves into small, modern, competent and competitive high-tech and R&D based enterprises. Such companies will be very important for the economies of their states, but they might be also very important for economic integration of those states in a new global digitized economy, integration into the EU, integration into a common European defence industry, etc. Such companies may also function as an interface between their transitional economies and economies of well-developed states, enabling diffusion of knowledge, technologies and modern type of business management into their states. Without such successfully transformed SME's, there is no compatibility and interoperability of these states with the NATO defence industry business sector and there is no tighter integration within the NATO structures.

1. Offsets and the New Specific Role of Defence SME's in Small Transitional States

There are many chances and opportunities for new specific positions and roles which defence oriented SME's may take in the near future in the small transitional countries. One among many may be related to the role of SME's in offset policy implementations and regulations. Implementation of offset policies in many transitional states should be very attractive for many small and medium-sized enterprises, as it is in the majority of developed states. In today's defence and civilian markets, offsets will have more and more acceptance and importance and this trend is not likely to be reversed in the near future. Creative, innovative and smart offsets techniques require specialized SME's for consulting assistance, education and training for intelligent and effective compensatory management transactions. It may be a challenging business opportunity for many consulting and management-oriented emerging SME's, which mainly do not exist yet in the majority of transitional developing states. A wide spectrum of activities contained in offset agreements related to the co-production and subcontracting, technology transfers or licencing of technology, in-country counter-procurements, marketing assistance, financial assistance, investments, joint ventures, etc., offers many new attractive and challenging additional opportunities for existing and emerging SME's. Offsets can bring jobs, technology and production experience to many transitional states and SME's, creating and maintaining a new more flexible, competent and competitive domestic industrial base and defence technologies. Offset programs can be very important for restructuring and developing of national economies in many transitional states, developing their defence industrial potential, opening new cooperation, organizational improvement of their enterprises, development of their scientific research institutions, better higher education, as well as restructuring of their regions threatened with unemployment. Some transitional states may also use indirect defence offset procurement as a tool for development of their civilian industry, such as in tourism, agriculture, the food industry, building their infrastructure, etc.

But many transitional states are not yet fully aware of the importance and potential of compensatory transactions in military procurements for their economic recovery and restructuring.

One specific role, which many defence-oriented SME's will have more and more of in the future in small transitional states, is the relation to the maintenance of their newly-procured military equipment. The main weapon systems and sub-systems in these countries will be procured in the future, mainly from developed industrial countries, due to limitations and inabilities of small transitional states to design and to develop modern sophisticated state-of-the-art weapon systems. Therefore, maintenance of their military equipment which needs servicing and repairing, even in some cases in the next 20-30 years, as a contractual agreement of direct offset arrangements, might be a very predictable and reliable long term business for many SME's in these states. The specific role of SME's in maintenance of these systems and sub-systems might be very challenging in a technical and technological sense as well as in many other aspects. General transformation of defence industries in small transitional states from the role of prime supplier into the role of maintenance provider is one of the guidelines for SME restructuring and their transformation mainly to maintenance provider. These facts bring us closely to a situation where small transitional states as buyers of complex defence equipment will function almost exclusively as maintenance providers based on their domestic SME's.

And finally, apart from a very few SME's which will function as defence-oriented Think Tanks in the areas of defence policy, defence industry conversion, SME restructuring and transformation, new type of PPBS strategy for a time of heavy defence budget downsizing etc., there is no competent public domain discussion related to many relevant strategic defence issues for these transitional states. There is no smart and competent political decision with long-term visions. There is no successful transformation of their SME's and, finally, there is no successful transformation of their states.

2. Some Guidelines for Transformation of Defence-Oriented SME's in Small Transitional States

Successful transformation of former defence-oriented SME's may provide new power and capacity in economic consolidation and economic recovery of small transitional countries and may have an important role in future international defence industrial cooperation. What do existing and emerging defence SME's need to know in order to cope with reduction of defence markets and with the increase of emerging civilian markets? What are the guidelines for transformation and restructuring of these small and medium-size transitional enterprises? What is the role of government? These are only a few of many important questions which require a detailed and careful analysis and appropriate answers. Some general remarks and guidelines regarding their successful transformation will therefore be outlined.

2.1 Organizational and Business Culture

Organizational and business culture of SME's is very important for their ability to be competitive and to be marketably self–adaptive in a changing world and a changing business environment. An organization where people can expand their intellectual capacities, where people can create the results which they truly desire, where new creative and innovative ways of thinking are nurtured, where individual and collective aspirations and objectives are harmonized, where the culture of innovation and creativity can flourish, where people are continually learning how to work and learn together, might be a successful organization. Unique team vision, team cohesion and team discipline, even productive conflicts of ideas rather than conflicts of rigid individuals and rigid policies are important for a new modern type of organizational culture. Experience shows that talented engineers appreciate discipline when it is implemented in a way that makes sense for the particular company and project to help them to succeed. Identification of emerging long term stable civilian markets and customers, diversification and balancing among defence and civilian services and products, shaping and education of potential strategic customers, more innovative management and marketing, creative leadership, organizational flexibility and its fast adaptation to the rapid market changes and environment, flexible hiring and contracting of manpower, an innovative and modern type of management, minimization of bureaucracy, a level organizational structure, etc., would be very important characteristics of an SME's new business reshaping strategy.

2.2 Personnel Management

Small companies live and die on the engineering talent they are able to hire. Talent alone is not enough, but it is very essential. Small companies which wish to remain competitive on the global market, must find ways of attracting and retaining their talents. The battle for talent must be a crucial element of their new modern agile business strategy and the minimization of friction between talented individuals must be part of their market strategy survival. Certainly companies with a small staff must take great care to make the best use of what their current employees are capable of doing and what they want to do. In such uncertain and turbulent times, characterized by heavy defence budget reductions, companies with small staff are in a surprisingly stronger position in comparison to medium size and large enterprises. Tighter connection and cooperation with universities as a main source of highly educated and flexible manpower, part-time jobs and project-based hiring and contracting of manpower, are some elements of the smart self adaptive personnel SME strategy as a mechanism for avoiding financial risks with company personnel and the maximization of R&D and business potential.

2.3 Research and Development

Generally, competitiveness of defence-oriented SME's in the small transitional states in globalized markets will depend on their effort to encourage their internal innovation practice, to increase applied R&D activities, to establish and to increase cooperation within state R&D institutions and to put more R&D based innovations into their products and services. But only a very few of former defence-oriented SME's in transitional states support R&D or outsource this as a part of their innovation practice. Therefore, the governments of these transitional states must take a proactive approach to scientific and technology promoters, stimulating applied R&D-based innovation of products and services in these SME's. Firstly, they can encourage long-term relations between R&D scientific community and research institutions with SME's. However, the problem in developing transitional countries, like Croatia, is in the fact that R&D performed at government sponsored scientific institutions and universities is mainly not applicable and is too far from any serious market commercialization. On the other hand, the number of SME's in small transitional countries which have advanced technological capabilities and which may have an interest for applied R&D cooperation is very limited. The main reason for such a situation is a lack of entrepreneurship skills and competencies in state-sponsored research institutions, a lack of financial support for application-oriented R&D projects, the level of technical competencies of the SME's management staff and sometimes the level of technical sophistication of their products and services. Researchers in state-sponsored scientific institutions enjoy their "safe haven" surroundings. Through a fear of living in the market, they do not want to taste the experience of the market's risks. Without integrated efforts of private SME's and state owned R&D institutions and commercialization of their joint products and services, there are no contributions to the overall technological in-state expertise, economic growth and new employment. The culture of evaluation of R&D only through the citation index, without an evaluation of the commercial free market in technical cooperation with SME's, will be a big failure for many small transitional states.

This means that successful and intelligent transformation of these defence SME's and their consolidation is not possible without smart government support and government funding. But government authorities must want and must know how to change the innovation practices in these SME's and how to improve their technological competitiveness. Smart strategy of changes and the innovation practice must be embedded in organizational routines of these SME's, by appropriate training and education of their leadership and management staff. However, the remaining communist mentality and culture in these states and their enterprises are the main obstacles to their transformation. Furthermore, we have to take into account that marketing of applicable R&D in small transitional states, like Croatia, is extremely difficult and barricades against its acceptance are very high. Management staff in the majority of these SME's is not yet fully aware that innovation based on R&D is better and more profitable than the old way of doing business. In a cultural sense, many SME's are cautious, suspicious of any change, because they hesitate to leave the old Cold War proven ways of doing business without innovation. A long-term contact between scientists and SME's can initiate a positive interaction which can result in new projects and innovation. A new type of doing business, based on dual use technologies, like software development and off-the-shelf hardware, might be very profitable. Some of the small transitional countries may also enter into the forefront of technology development, but the prerequisite is much higher R&D investment, at least 3% of their GDP annually into applied R&D. In addition, what is very important to stress and to anticipate is a significant reduction of conventional defence technology, as well as the advancing modern advanced defence technology market. In the past, technology development was pushed forward by military demands and military technology used to lead to civilian spin-offs. But today, civilian technology leads to military spin-offs. As an illustration, in the semi-conductor business 30 years ago the defence sector had more than 90% of the market, while today the defence sector has about 0.3% of the market.

2.4 Financial management

A smaller and smaller MOD budget has a direct negative impact on its structure for many reasons. Ideally, an MOD budget is based on the balance between total personnel income, acquisition and modernization programme costs and operational expenditures. However, in a period of heavy budget downsizing, due to social stability concerns personnel payments can cover more than 70% of the MOD budget. This MOD budget deficiency has a direct negative influence on acquisition and modernization programmes as well as in the state defence industry business. Without MOD R&D funds, lack of experience and familiarity with new high-technologies, lack of society's confidence in researchers and a remaining communist culture and mentality which presumes that investment in private enterprises leads only to corruption and criminal activities, it makes it very difficult to take a significant step forward toward successful transformation of these SME's in many transitional states. Without state supporting programmes for defence SME restructuring in the form of appropriate loans and grants, their transformation in a majority of transitional states will remain very uncertain. With a lack of management skills and some other business risks, many SME's in a time of very uncertain MOD budget size and structure are not willing to take loans. Without the MOD, as a reliable and predictable customer and partner, there is no guarantee for any SME's that they will succeed and will have continuing revenues for loan rate payments. So it can be concluded that a loan is inappropriate, in an uncertain time and an uncertain environment, as the only means for a SME's successful business.

Venture capital still does not yet exist in developing transitional countries. The reason is in underdeveloped financial markets in these countries. Additional development of investment and pension funds, reform of the health care system and diversification of the insurance market, will lead to new trends in the financial markets of these countries. Hopefully, very soon in these states, new investors will seek for new opportunities to invest their funds into high return and high-risk assets. Venture capital funds could be one of the appropriate solutions. An excess of venture capital, for example in the USA, with an amount of more than $40 billion, will also seek for new opportunities in emerging markets. This could be a chance for the development of venture capital markets in these countries. But a fact, which has to be considered, is that small entrepreneurs in transitional states are mostly sceptical regarding venture capital and initially they reject any idea that has any connection with it because they are not willing to lose full control over their companies by giving or selling a part of company shares to venture capitalists. With such business logic, they are missing the chance for rapid growth and development of their companies. Better promotion and marketing of venture capital and promotion of the benefits which this can bring to SME's, may reduce entrepreneurs' rejection of venture capitalists. This kind of marketing could be done by the state agency for investment promotion and should be focused on the small and medium entrepreneurs.

In developed countries classic venture capitalists are investing into whole companies, increasing their values and exiting from them. But cooperation between entrepreneurs and venture capitalists could be based on specific projects and could be very attractive. Instead of investing into a company's equity, venture capitalists might invest funds and provide consulting services to the company in order to develop specific new products and services. The venture capitalist will benefit from a percentage of sold products or services in the future.

There is no doubt that transitional defence-oriented SME's, embedding some of the guidelines and recommendations outlined in this section may have a better chance of survival in the turbulent transitional time of defence market shrinking and restructuring.

3. Croatian Case Study

The case study described illustrates the programme strategy transformation of one former Croatian, mainly defence-oriented small enterprise, which is trying to survive in a time of heavy defence budget downsizing by making a balance in its defence and civilian programmes. This transformation and evolution is based on product diversification and transformation of the old business of defence-oriented strategy and products to the new more civilian markets focused on business strategy and products.

3.1 Old business strategy and products

The old business strategy and products were related to modernization and upgrading of some weapon systems for the Croatian Armed Forces with the objective of increasing their lifecycle by replacing an old analogue electronic sub-system with a new commercial state-of-the-art, off-the-shelf digital technology as well as improving their technical and tactical features including design and development of some completely new products. Some of these efforts and products were related to the modernization of the old analogue guidance unit for AT-2 and AT-4 guided missile anti-tank systems with state-of-the-art digital technology in order to reach greater maintainability and provide some new technical and tactical features, like improved IR CCM capability or improved hit probability of moving targets. This also included design and development of new maintenance and troubleshooting equipment for AT-2 and AT-4 missile systems which enable automatic testing of launcher and missile parameters; design and development of new sophisticated multi-level Hardware-In-the-Loop (HIL) simulators as a powerful infrastructure for modernization of guidance and control hardware and algorithms; design and development of PC-based training simulators for AT guided missile gunners based on a high performance multiprocessor system which provides facilities for a complex and highly realistic simulation of the real AT systems, such as a 3D terrain map for the tactical scenery, 2.5D photographic imagery of real landscape or artificial 3D scene, 6DOF model of the missile's dynamics, 3D model of the target kinematics, different terrain types, different targets and trajectories and different meteorological conditions; design and development of an air-defence, state-of-the-art simulator based on virtual reality technology for the man-portable air defence system "Igla", etc.

3.2 New business strategy

The new business strategy is trying to protect and to adopt company core competencies and expertise through more civilian-oriented and more profitable commercial projects which offer new business opportunities such as the Whiplash Injury Integrated High-Tech Diagnostic System (WI^{2}HTDS) project, primarily intended for assessment and classification of whiplash injuries, including rehabilitation and quickening of the recovery process of the injured. The system is based on virtual reality technology and biofeedback sensors and it immerses a client in a completely new virtual world, which alters his impression of the real head-neck range of motion. Virtual reality, biofeedback and appropriate digital signal processing algorithms and complex statistical aggregated correlation analysis detects the inconsistencies in the client's head-neck range of motion, thereby reducing the chances of faking an injury. From the results of the analysis, conclusions can be drawn, which can be used to create a complete and objective clinical profile of the client, thus enabling financial compensations to those who really need them and preventing unnecessary expenditures. Humanitarian high-tech multidisciplinary projects based on technologies of virtual reality and digital biofeedback related to the therapy of the Croatian war veterans who suffer from PTSD is also an example of a civilian-oriented project. The main objective of this project is related to the design and development of complex dynamically controlled biofeedback- based adaptive scenarios, which have the objective of relieving the therapist who guides the patient through a virtual reality scenario from routine tasks, thereby developing an expert system that could automatically select and adapt therapeutic scenarios based on the patient's current level of emotional excitement. MOD projects related to the psychological assessment and selection of combat-ready gunners based on biofeedback measurements of their emotional characteristics and profile under stress, use similar dual-oriented technologies and expertise.

4. Conclusions

Competent and smart defence policy, competent and knowledgeable MOD decision makers, their appropriate training and education, think tank assistance in design and development of defence strategy, are some of the most important prerequisites for smart and predictable defence budget and defence industry business in these small transitional states. On the international scene, cooperation between NATO and Partner transitional nations can make a significant contribution to the process of successful defence SME transformation as well as their adaptation to the new reality, characterized by defence market shrinking and expansion of newly emerging civilian markets. A proposal for NATO assistance to the former defence SME's in the transitional states could be related to the selection of a very few defence SME's in each Partner transitional state by assisting them through workshops, seminars, conferences and symposiums to speed-up the process of SME adaptation to the new business and security conditions and environment. The topics of these seminars and workshops could be related to modern organizational structure and management; offsets and SME's; personnel management; venture capital funds, etc. The foundation of a NATO Venture Capital Fund for investment and assistance of former defence SME's in a small transitional state as a leverage for its transformation and adaptation to the NATO defence industry business standards and procedures, might also be a very attractive proposal.

Defense Related SME's: Analysis and Description of Current Conditions
F.D. Carvalho (Ed.)
IOS Press, 2004

Small and Medium Enterprises in Central and Eastern Europe: from Absence to Emerging Dynamics

- How to effect international support projects successfully -

Heinz FIEDLER
President SPICE Group, Germany

Abstract. This paper briefly describes a few indicators for SME development in Central and Eastern Europe and elaborates on basic factors (mentalities, skills, markets) to be taken care of in running successful economic development projects with a special focus on matters of technology transfer and technology business incubation. Furthermore, some key factors (local support, integration, perspective) for project success and tools to be used are discussed.

Emphasis is put on the "human component" and "networking" as basic requirements for reaching continuous and sustainable changes. Experience from the SPICE Group network and projects carried out in this framework are exemplified.

The presentation focuses on outlining some experience generated from concrete projects and in trying to generalize such experience to come to some "recommendations" for undertaking international project work. The main component of these recommendations is the "human factor". Persons, not organizations, act in networks. People, not institutions, do something. Therefore it is only logical that education and training are the key issues for all activities in reshaping and (re-)developing economies and societies. The aims are to support people in setting up new initiatives and to enable them to further develop such projects; to create new role models and support "multiplying" effects.

SME's in Central and Eastern Europe have developed from absence to becoming a driving force for the (economic) development of the nations. SME's did so despite a still, at least to a large extent, hostile environment because they are driven by people, by entrepreneurs. International support projects, when targeted at people and infrastructure, help to make this possible

Introduction

Small and medium enterprises did not exist in Central and Eastern Europe just a little more than a decade ago. Today they are significantly contributing to the national economies of this region. Obviously there are important differences between the countries of Central and Eastern Europe. Due to political developments, e.g., (potential) NATO or EU membership and national influences, the SME sector has developed to different levels. In Central Europe (Czech Republic, Hungary, Poland, Slovakia) as well as in the Baltic States (Estonia, Latvia, Lithuania) SME's have become an important economic factor – certainly with much potential for development and many needs for adaptation of rules and regulations. In South Eastern and Eastern Europe, however, the legal and fiscal framework still needs to be adapted to enable, encourage and support the establishment of new and the development of existing small companies. Taking into account that ten years ago not even the concept of entrepreneurship was known in the former "Eastern Bloc" the development since 1990 has been dramatic.

Restructuring of the national economies has taken place in just a decade to an unexpectedly high level, as some might say. But by far not sufficiently, as others might state. However, keeping in mind that in some cases even the formation and structuring of a nation had to be achieved first before economic development could begin, the status reached is more than realistically could be expected in a little more than ten years' time. Naturally nations having an "industrial history" could develop from an adventurous basis. This is an important reason why the Central European region and the Baltic States developed comparatively faster than other countries. Another reason lies in different political strategies (possibly better to say developments). While some countries quite quickly laid the basis for business development by reinforcing "old" laws and regulations (even knowing that this could not be the final optimum) and by pushing the privatization process forward, others, even today, do not have the basic legal framework in place.

Military downsizing and the resulting conversion of industrial production is just one part of the problem. There is not too much of a difference between the task of restructuring a "socialist production conglomerate" to make it a market-driven industrial enterprise and the task in reaching the same for a military production plant. Because of the size of the former "socialist production conglomerates", the dimension of this problem (closing or restructuring) is similar to base closure problems. The basic requirements are similar in both cases. Certainly the military sector has some specialties; however, many basic problems are similar and are related to "society conversion" rather than specifically military conversion. Corruption, of course, is another important factor very much influencing business development. Also, in this respect, the countries of Central and Eastern Europe have developed very differently. But in this report the general political and social conditions at large should not be discussed. Emphasis is put on experience from concrete, small projects and requirements for running such projects successfully.

1. Basic Factors

Much has been said about the most important basic requirement: a stable, reliable and favourable legal and fiscal framework. However this topic should perhaps be set aside as influencing this factor means "big policy making". In addition to the fundamental political and regulatory framework, the most important topics to consider are mentalities, skills and markets. These factors, to a certain extent, can be influenced positively by (a large number of) "small" projects.

1.1 Mentalities

This factor is not only important for the former "socialist" societies. Differences in habits, traditions and mentalities have been much discussed as have the "entrepreneurial cultures" in the US and Europe. Such differences do not mean anything good or bad "per se". Differences can have positive impacts on entrepreneurial or business developments. To know this means to carefully adapt whatever experience to a specific environment and to local needs.

Having been raised in a centrally planned, or centrally commanded military, environment, how can you understand, how can you feel and live the concept of individual responsibility, the concept of entrepreneurship? Here, speaking about *Central* Europe, you possibly had a grandfather who was an entrepreneur or a manager of an industrial enterprise. Something to begin with. However in other regions where market oriented industrial development had never taken place, whom to refer to? What are the values? What are the role models? The structure of the society in general is another factor influencing mentalities. There are traditional family or "clan" oriented societies (not only in Central and Eastern Europe) making it more difficult to introduce the ideas of individual responsibility and entrepreneurship. The head of the "clan" is the decision taker – not the young, well educated, motivated would-be entrepreneur. To change mentalities needs time because it means changing human behaviour. However time is not all. If the environment does not change, or if there is no hope at least for change and if no new "role models" are developed, time will change nothing.

Mentalities, cultures, habits, education etc., are also reasons for the differences between the economies in the industrialized world, whether it be Europe, Asia or America. Despite "globalization", cultural heritage influences the behaviour of people, the way they live together and the way they do business. And cultural heritage is an integral, important component for the functioning and development of economies and societies. This only reminds us of the fact that there is not a single model for "how to do it" successfully. The task remains to find the individual way.

1.2 Skills

Having been educated to follow instructions given by "someone else", whether it be by the family (clan) head, the political leadership or a military commander, how could you have the skills, the tools necessary "to do something on your own"? How could you write a business plan for a company - or for your life? This means the basic skill of managing your own life is often still missing. Certainly this matter relates to mentalities, but the necessary skills and training in their use can be provided. In addition to this very fundamental requirement there is a wide range of instruments and tools needed to run a business, manage an administration or to do research projects. Therefore training was and remains one of the most important matters to deal with in supporting the change of societies. Much has been done in this field. But by far not enough. Often "training" was and still is understood to be too "instrumental". It is not sufficient to provide tools – instructions in how to use them are needed as well. This leads to the next problem. Many tools have not been adapted to a different environment. Many trainers were focused too much on their home experience, not caring sufficiently about the fact that what works at home may not necessarily be appropriate elsewhere. Furthermore, societies and economies in Central and Eastern Europe have developed. Training offers today have to be different from what they were a decade ago. And all training has to take into account that there are cultural differences which have to be accepted and integrated. The goal cannot be to "make it the same way" everywhere – ways have to be found for using the local potential in the best possible way. And this might be different to anywhere else in the world. However, this way of "adaptive training" requires much more openness, information and preparation than simply doing the same training seminar in America, Western Europe or Asia. More often listening rather than talking is needed.

1.3 Markets

Having grown up in an environment where the government decided how you lived, what work you did and where, what you could buy – if you could buy anything at all, how could you understand the mechanisms of demand and supply in a market driven economy? It is easy to understand that the first business concepts implemented by individuals in Central and Eastern Europe were simply trade or supply-oriented; just offer anything – people will buy it. Why take the risk of investing in production facilities, if fast and easy trade can make a profit?

In Western industrialized countries technology/knowledge-based SME's are part of a diversified, sophisticated supply chain. And even in this environment individual small, specialized companies find it difficult to convince potential large customers that they can reliably deliver. This is why small business networks (or stronger forms of cooperation) have been developed to approach such large enterprises with a more complete offer and to create a better image.

In economies in transition, SME's find themselves as "stand-alone" activities. There are very few if any large functioning enterprises as customers. There is no governmental (military or research) budget as one of the driving financiers for innovative approaches. There is little or no domestic demand for (expensive) "high tech" products. Export as an alternative? Possibly, but still a difficult task. In addition to the problems at home of insecurity, mentalities and skills, the foreign market is another completely unknown place. It is no surprise that in Central and Eastern Europe the technology/knowledge-based business sector is even more difficult to develop than it is in industrialized countries.

Before discussing some approaches to solving the problems described, let us have a brief look at the actual situation.

2. The situation today

"From zero to a respectable level" could be said for the development of the SME sector in Central and Eastern Europe. The following figures are not intended to provide a scientific comparison. There are still too many differences in definitions, understanding and statistics. For example the term "Small and Medium Enterprise" does not mean the same in every country. And there is a large gap between "registered" and "active" companies. But the few figures indicate the actual situation.

The following small overview, irrespective of "statistical misinterpretations" which are certainly involved, shows that much has been achieved during recent years. For comparison, data from Germany is listed as well – a country not known as the "cradle of entrepreneurship". At a first look, the level of SME contribution to GNP in Central and Eastern Europe is not too far from "Western standards". But the numbers only show SME's in general – including trade business which still is the main part of SME activities. Definitions of what is an SME differ. Statistical methods and standards are not the same. For example in Germany only companies registered for taxation are counted. This means a large number of so-called "micro enterprises" do not appear in the statistics. A more detailed analysis of the type of companies in Central and Eastern Europe and specifically looking at manufacturing and/or technology/knowledge-based firms would show a less favourable picture. There still are important obstacles to overcome. The main factors have already been outlined above; the legal and political framework, mentalities, skills and markets. But, and this is all this table intends to illustrate; there has been much development in the SME sector in Central and Eastern Europe.

SME's in Central and Eastern Europe – some Examples					
Country	**Population (million)**	**SME size Employees ***	**Number of SME regist-ered/working**	**Portion of GNP produced by SMEs**	**Inhabitants per Company**
Bulgaria	8,3	< 250 **	928,451 / 225,648	29,5 %	37
Estonia	2,4	< 250	31,346	55 %	76
Hungary	10,0	< 250	800,000	40 – 45 %	12
Latvia	3,8	< 250	37,980	53 %	100
Lithuania	3,7	< 50	54,134	48 %	68
Poland	38,7	< 250	3,000,000 / 2,000,000	55 %	19
Romania	22,4	< 250	630,000	55 %	35
*EU standard: micro: < 10, small: < 50, medium: <250 ** in Bulgaria statistics define SMEs as <100, but here we use the number for <250					
To compare: Germany	82,1	< 500	3,200,00	57 %	25

The importance of the SME sector for a national economy is not questioned in general. For example, in Germany politicians repeatedly stress the importance of SME's because they

- represent 99.3 per cent of all taxed legal entities
- "produce" 44 per cent of all taxable turnover
- make 46 per cent of all investments
- provide 69.3 per cent of all jobs
- provide 80 per cent of industrial training

And in saying this, the same politicians increase the administration and taxation burden on the shoulders of SME's. Similar statements could be made for (almost) every country. Obviously there are differences between recognizing the importance of the SME sector and translating this knowledge to more than "verbal" support for SME's. It seems this topic will remain on the agenda everywhere and forever.

Many institutions work in different kinds of programmes and projects to develop SME's wherever in the world. However, the result of such activities depends on the political framework; certainly the most important basic factor. This factor can hardly be significantly influenced by activities undertaken by people or institutions attending this workshop. Therefore it does not make too much sense to repeatedly present the differences, problems and needs in this field. It seems more interesting to look at what (a large number of) concrete activities can achieve in restructuring the economy from a "bottom-up" approach. Such projects are supported by many different national or international governments, organizations or institutions and for different reasons. They have in common that they want to support the development of the economy in different countries and to support the development of their own economy.

One remark based on concrete experience is important; obviously, although desirable, it is not possible to coordinate such activities at a "high" level. National programmes are made to reach nationally defined goals. International organizations have their own set of aims. No organization wants to share success (or wants to be blamed for someone else's failure). But there is an endless number of cases where different institutions want to do the same thing in the same place with the same partners in the target country at (almost) the same time. It would make sense to try to find some possibilities for coordination at the "field work level". This is not an easy task. But the better the information system between those who do the projects, the more likely are the possibilities for mutual benefit. After this idealistic remark, let us come back to the real world.

Within the network of the SPICE Group and related partners there is a very wide set of experience gained from projects which train people, which help them build business incubators and technology parks, which support the development of a local/regional economic development strategy – and more. What follows is an outline of some aspects of experience gained.

3. Key factors for successful projects

This collection of factors is not based on extensive scientific studies; it refers only to direct experience and exchange of information between "doers".

3.1 Local support

This is the first and most important requirement. What is meant is not "support as long as finance (from abroad) lasts". This kind of local support is easy to achieve. The difficult part is to identify "local players" who are willing to commit themselves to the aims of the project and are ready to continue activities with a long-term perspective – even after the project with foreign partners and support is over. Of course, it is not an easy task to find such local players before a project begins. Often you know who would have been the best partner after the project is finished. The necessary basis for finding the best partner is knowledge of the local situation and people. This knowledge normally only can be acquired by integrating people in the project team who already have this experience or by longer term cooperation. However, the term "cooperation" does not necessarily mean that long-term projects have to be the basis. Continued contact, even without a concrete project or financial component, can secure the basis for a new project.

3.2 Integration

To integrate means to truly accept local partners into the project work and team and at the same time to support the different local players in finding a common ground and beginning to seriously work together.

Project work integration means listening to information and views provided by local partners and making them personally, and their advice, an integral part of the project activities. If there ever was a time when Western experts knew better, this time is now definitively over. But there are still too many "we know how to do it" or "just do it as we do it" approaches which ignore local habits, conditions or interests. Naturally as long as foreign money flows, local partners will more or less follow the "advice" of such experts. But as soon as the project is over just one more study ends on the shelves collecting dust. There are endless numbers of projects sharing this fate. And financiers of such projects keep wondering what has happened to the money and why the so-promising project does not continue.

Support cooperation between local players is important for several reasons. "Solidarity" was the word from former times. "Competition" is the new key word. But what does competition mean? Is it "alone against the rest of the world"? Or is it forming, perhaps temporary, alliances to share information, combine resources and skills to gain a competitive advantage or a stronger voice or more influence? To illustrate the importance of this question, here is an example. During the first years of work in Central and Eastern Europe we were wondering why business incubators and technology parks in these countries were hesitant to form associations. The reasons were simple; firstly it was expected that the government would do something in this direction. And in some cases the government did. Looking back we can say it was with "limited success". Secondly, and this was the more important reason, the understanding was that business incubators and technology parks compete, e.g., for the same sources of finance. Therefore it simply appeared to be against their own interests to share information or to cooperate.

3.3 Perspective

In by far the majority of cases, projects are run with just one aim; to comply with the terms of reference, to deliver the final report, to cash in the payment and to get out of the project again as soon as possible. However, projects aimed at supporting economic development and entrepreneurship are by nature medium and long-term projects. This is even more important in the situation and environment of Central and Eastern Europe than in Western countries. Longer term commitment is needed because it builds up the necessary information basis, understanding and trust on both sides. Economic development projects including business incubators, technology parks or technology transfer initiatives will show results only after several years. This means they have to be organized in a way which will make them, as far as possible, independent of political influence. They need a long development period and therefore require long-term cooperation, but this is not easy to finance within the patterns of usual projects.

An example of how this worked comparatively well was the establishment of business and innovation centres in Eastern Germany. The kick-off was a federal government programme supporting the first 15 such projects. In addition to a number of "normal" conditions, such as providing a business plan for the project, a specific requirement was to find an existing innovation centre in Western Germany as a partner. For the first few years this partnership was financially supported, but with decreasing amounts. Now looking back at the experience gained during that time, it can be stated that all projects (and meanwhile more) have developed successfully. In almost all cases the links between the project partners are still maintained and some project partnerships have become a business cooperation. To some (but not to a sufficient) extent, this experience has been later transferred to German support programmes for innovation centres in Central and Eastern Europe.

3.4 Cooperation

Having described the cooperation between "Western" and "Eastern" partners, let us now have a brief look at what could or should be done in "West-West" relations. If financing institutions cannot/do not want to cooperate, the "field workers" have to find ways to make it possible – at least occasionally. It is easy to say, but not easy to do. An informal network such as the SPICE Group can help to provide information about *planned* project activities. Emphasis has to be put on early information. Once a project is up and running it most likely will be too late to coordinate with similar or supporting activities of other projects. To inform other, potential competitors, about "projects in the pipeline" needs much trust; a factor which can only be developed over a period of time. However, once such cooperation is achieved, the benefits are obvious for all partners involved.

4. Tools

Having described the basic requirements for successful projects, some tools which need to be implemented and developed in Central and Eastern Europe will be discussed. The underlying idea again is not just to deliver such tools, but to enter into a cooperation aimed at supporting the understanding (on either side) and the adaptation and integration into an environment necessarily different to where the tools have been developed.

4.1 Build a Basis

One of the key issues in providing fertile ground for business development – again in addition to the already-mentioned basic framework conditions – is to establish a broad and solid acceptance of the goals and tasks. Economic development (restructuring, revitalization) projects need a wide basis of supporting institutions – or, it might be better to say, supporting people. The time and resources invested in establishing such a basis will pay off later in the project and, even more importantly, in the continuation of activities in "normal life" after the project is over.

The tool to develop such a support basis is simple to name, but difficult to do; build trust and mutual interest. Meet with people, explain what you want to achieve and find out if your partners understand what you are talking about. Often enough the same terms are used, but with a different understanding of their meaning. Take a simple example; the word "innovation". We all use it. We all think we have a common understanding of this term. But some understand it as the process leading to a new product, service or process – others see it as the result of this process. Innovation may mean product-related – or you may think of this term as a change in society. Or what is an "entrepreneur"? Is it the yuppie type of person; is it the well-educated young person in "Nikes" with a brilliant technology-based product idea; is it the rich guy driving a Mercedes and smoking Havanas sitting in his tailor-made suit behind a huge and empty desk; or is it the hard-working, risk-taking enthusiast realizing his idea of a better service or product? Imaginations are different. Interpretations as well, even when we talk about people in a community somewhere in the industrialized world. How many more differences must there be in expectations, interpretations or imaginations if "West meets East"?

More than a decade after "the Wall came down", we want to believe that a common understanding of basic ideas and definitions has developed. Not sufficiently! And the "rules of the game" are not the same, not even in what we call "market economies". Therefore, building a basis of mutual understanding for every project, is a key issue.

4.2 Offer a Toolbox

The most important driving force for business is innovation and the continuous search for and implementation of new products and new services to survive in a competitive ("globalized") environment. To support this process, the partners (such as business, research, administration) with different roles, functions and tasks in a society have to care about links, have to keep their doors open, have to network. Besides the basic mental readiness, tools have to be at hand to encourage and support the flow of information and cooperation between such partners; partners who have different aims, tasks, legal structures or motivation.

Behind the term "technology transfer" you can find a wide range of tools for this purpose. This is not the place to describe them in detail. There are many different instruments available – from (scientific) publications to active selling initiatives, from educational programmes to personnel placement, from seminars to "spin-off" promotion. These instruments offer a wide range of possibilities and require specific resources. Much experience with such tools is available throughout the world. They have different effects in different environments. Therefore it is best to provide information about the tools available and how they might be used. This will enable the local partner to use whatever tool whenever it is needed.

Another set of tools is found behind the term "business incubator". This is not an entirely different set of tools. There are some technology transfer instruments in this toolbox. A business incubator seems to be very simple at a first look; erect or renovate a building with office and production space, establish a service unit for secretarial, administrative and advisory services, and then let the tenants in. However, this is only looking at "hardware". What makes a business incubator work is the "software", the networking and the know-how. Once you start an incubator you find that the early phases (building, set up services) are comparatively easy. The only difficult part is acquiring the finance. Even finding the first set of tenants – looking back in time – has been easy. The business incubator was new, received much attention and was attractive. What makes the difference is keeping the business incubator developing, guiding existing tenants to leave the incubator – but keeping them in the network. Adapt your services to the changing needs of your clients. Attract new tenants with an eye on what are "future technologies. All this can only be accomplished with the right understanding of the processes and knowledge of the necessary tools – from motivation to information, from inspiration to planning, from leading to balancing, from financing to accounting, from high technology to housekeeping. The spectrum of required knowledge is wide. And the most important component is - networking

No "models" are available for economic development projects. In speaking about international support projects for such initiatives, the goal must be to provide a "toolbox" for the local partners, giving them instructions for its use and enabling them to find and use the right tool whenever it is needed.

4.3 Build a Network

If the first two components are in place there already is a local network. But this needs to be linked regionally, nationally and internationally:

- because the clientele you are dealing with may need the international component – for example, if you speak about a "high-tech business incubator" or "technology business clusters"
- because the local people who continue the normal work in what was "the project" will need further information, contacts, support or finance
- because the knowledge and experience gained from this project should be made available to others facing similar problems
- because you, for your own interest, may want to make use of what you helped to build.

To build such a network is one thing. Sometimes at least it is part of the financed project. But to build a network in a way that it continues working after the project finance is finished is a different task. The most important factor for reaching this goal is the personnel component. The second factor is to think about the "software" and not the "hardware". In other words, it is simple to make something happen as long as someone pays for it. What makes it successful is finding out what the interests of the network partners are and to build and develop the network in a way that the participants benefit from investing in the network. "Investing" here does not only mean to pay fees, but to actively work in the network. To achieve this is not a one-time effort. It needs continuous, never-ending work.

5. Illustrations

To illustrate the meaning of the afore-mentioned key factors, such as cooperation patterns and tools, some aspects of concrete projects will be described. These examples are based on a project carried out for the German Federal Ministry for Economy and Technology by ADT, the Association of German Business Incubation and Technology Centres and the SPICE Group.

The general aim of the project is to support innovative German SME's to enter into foreign markets. A set of instruments is used to reach this goal, e.g., information, business-to-business meetings and – as a pilot project – specific training for intermediaries. There are many branch offices of German institutions in other countries. Therefore the pilot project was aimed at identifying and training qualified persons in the "target country" to support international business cooperation. A specific condition of the project was *not* to integrate participants into German institutions, but to let them continue working in their "mother" institutions. Without describing details of the project, the general scheme consisted of the following phases; identification of potential candidates, selection for a "screening" seminar, selection for an "advanced" seminar, internship in Germany (in SME's), concluding with an international workshop with the "best" participants and integration into a network.

This project began in May 2000 with the first seminar in Ukraine and concluded with a workshop in Berlin in May 2002. After a nationwide tender in Ukraine, a large number of interested people applied. The attraction certainly was not the training seminar. It was the possibility for the best three participants to go for a (paid) internship for (up to) three months in Germany. From these applications 25 were selected as participants for the first seminar. Consciously the selection covered very different educational and professional backgrounds. As a first "surprise", it turned out that these people were well-educated and highly-motivated. The trainers could learn something from the participants.

To give a general remark: this project time of two years was too short to reach all goals (dreams). Another example was that the financier did not have the patience to let the project develop. Several times during this working period the content, goals and work patterns were modified. But even though, there are some outcomes worth describing in the following two examples:

A young woman with excellent university education and several additional special business training programmes applied for the first seminar because she "did not know what else I could do" (no job offer). Meanwhile she attended the advanced seminar, took up two internships in Germany and now is working in Ukraine for a German company setting up their second branch office. In the first office she was an employee; now she is the person responsible for the project. But although this work certainly keeps her busy, she continues to be interested in maintaining contacts in the international network, providing information about her experience and learning from others.

A young photographer missed the "entry seminar" but was accepted for the advanced seminar because of his high motivation and engagement. Jobless, without too much hope of finding paid work in his home city, he first thought the seminar would provide a contact with foreign firms to work for. However during the seminar he focused his interest on how to start a business and how to use contacts with foreign business partners for this purpose. He began his first steps as an entrepreneur in a printing company after the seminar and before being invited for an internship in Germany. The result is the beginning of a cooperation between his company in Ukraine and a German printing firm.

6. Concluding Remarks

This presentation focused on outlining the experience generated from concrete projects, trying in general to describe some "recommendations" for successful international project work. The main component of these recommendations is the "human factor". Persons, not organizations, act in networks. People, not institutions, do something. Therefore it is only logical that education and training are the key issues for all activities in reshaping and re-developing economies and societies. The aims are to support people in setting up new initiatives and to enable them to further develop such projects; to create new role models and support "multiplying effects". To reach this aim, training does not have to mean preaching like a missionary. It means understanding the backgrounds, environments and needs of participants in training programmes (or partners in projects), developing the content needed for the training/project together with the participants/partners and to "live" what you teach. And reaching this aim needs time – something not sufficiently available in most international support programmes.

SME's in Central and Eastern Europe have developed from absence to becoming a driving force for the (economic) development of the nations. SME's did so despite, at least to a large extent, a still hostile environment, because they are driven by people, by entrepreneurs. International support projects, when targeted at people and infrastructure, help to make this possible.

Annex "What is the SPICE Group"

The SPICE Group
General Information

The SPICE Group is a global network of innovation centres representing **29 countries** including **17 national and international associations.** Members of the SPICE Group are experts in business incubation, innovation centres, technology parks, regional economic development, technology transfer and entrepreneurship. SPICE Group members represent different professional and organizational backgrounds and combine information, experience and skills from many different areas.

Aims and Tasks

The SPICE Group (ICECE Work Group) is a network of professionals from innovation centres (technology/science parks, business incubators) as well as industry, banks and government institutions, securing the best cooperation between *all* partners - in the private *and* the public sector - for economic and social development. The SPICE Group's work is based on mutual personal trust, free of political and institutional influence in an open, creative atmosphere.

The aim of the SPICE Group (ICECE Work Group) is to **increase knowledge** about business incubation, international business cooperation and promotion, economic policies and development and entrepreneurship as well as technology transfer by acquiring joint projects to :

• develop the **exchange of information** and experience in the fields of regional economic development, business and innovation centres as well as technology and science parks,	• develop **programmes for education and training** for managers of technology parks, business incubators, regional economic development agencies as well as small and medium companies
• promote the **establishment of networks or associations** (regional, national and international) and strengthen their cooperation	• **increase** the information and **understanding** of different cultures and "ways of life", especially for young professionals
• **transfer know how** and information in the field of business incubation and management of small and medium companies	• further **develop the informal international network** as a general link between existing or developing national, regional and international networks

SPICE Group Membership

Membership of the SPICE Group is personal, by invitation only, for lifetime and independent of the actual professional position. **Active personal engagement** in the global network is the most important criterion for membership in the SPICE Group. The first step to become a member, open to everyone interested, is the ICECE status. After a period of six months to two years of a mutually positive experience in cooperation, membership in the SPICE Group can be acquired.

Leading representatives of national associations of innovation centres (presidents, managing directors) can join without the trial period if they decide to commit themselves personally.

Principles of the SPICE Group

Members of the SPICE Group are personally committed to promote entrepreneurial activities as well as innovation in their countries – no matter which professional function they actually have and which organization or company they work for. They accept responsibility for helping the international exchange of information and cooperation as well as acting as disseminators in their countries of the collection and distribution of information.

The SPICE Group meets usually four times per year to exchange information and experiences as well as to prepare projects. If possible, such meetings are organized at the time of other events (conferences, exhibitions, seminars etc.) to ensure the best possible use of time and finance. Participation of members of the SPICE Group is based on their own resources. These meetings are for members and invited guests only. Public events of the SPICE Group ("ICECE Forum") are open to participation of all interested parties. Participation fees occur for non-members attending the meetings (as invited guests) and for public events.

The SPICE Group works on a **self-sustaining basis**. Members devote part of their resources (or resources acquired from third parties for this purpose) to support the activities of the SPICE Group and to support each other mutually in developing and organizing national and international events as well as acquiring funds for joint projects for promotion of innovation, business incubation and entrepreneurship.

History

Almost eleven years of active work began in Berlin in November 20, 1991 at a conference of ADT, the Association of German Technology and Incubation Centres. This was the first meeting of the "Innovation Centres in Eastern and Central Europe, ICECE Work Group". Participants decided to meet regularly and to establish this work group with the aim of helping to establish and develop technology parks and business incubators, to promote entrepreneurship and to support international collaboration of small and medium-sized companies.

To emphasize the development towards a **global network**, the name "**SPICE Group**" (Science Park and Innovation Centre Expert Group) was introduced at a meeting in Leipzig in Germany, 1997. In Spring 1998 in Poznan, at the first international science park conference organized in Poland, the SPICE Group held its first elections for the President and Board of Directors.

Results of SPICE Group Activities

- A reliable **network** of business incubation professionals in 29 countries
- **Publications**, such as a regular newsletter - the "ICECE News" and the **SPICA Directory**
  (Science Park and Innovation Centre Associations) or the series "ICECE Focus":

- Promotion of **collaboration** between ECE government bodies in the field of innovation and enterprise support
- Help to organize **international partnerships** and support for planning, establishing and developing business and innovation centres
- Support to establish **national organizations/associations** as well as preparing their local, regional or national meetings, seminars or workshops
- **Training** "on the job" between SPICE/ICECE members in various countries and by organizing international internships for "future leaders"

In almost **11 years of global networking** the SPICE Group has (co-)organized more than **80 events** (conferences, workshops, seminar training and exhibitions) **in 19 countries (**in Belarus, Bulgaria, Croatia, Czech Republic, Denmark, Estonia, Finland, France, Germany, Hungary, Kazakhstan, Latvia, Lithuania, Poland, Romania, Russia, Slovakia, Ukraine and the United States of America).
The SPICE Group has initiated the **first global directory** of Science Park and Innovation Centre Associations, the **SPICA Directory**, and at present is working on an Internet-based online version of this directory.

The SPICE Group has performed project work for national organizations, such as national Ministries and foundations or at the international level, e.g., for the German Marshal Fund of the United States.

Contact:

www.spice-group.net

The SPICE Group
(Secretariat Germany)
ZOOBOGEN, Hardenbergplatz 2,
10623 Berlin, Germany
Tel: +49-30-264 7070,
Fax, + 49-30-26470733,
e-mail: spice@spice-group.de

The Board of the SPICE Group
The elected Board of the SPICE Group consists of seven elected members: the President, two Vice Presidents and four Regional Directors.

President:	Heinz Fiedler, Germany
Vice President:	Janis Stabulnieks, Latvia
Vice President:	Krzysztof Zasiadly, Poland
Regional Director:	Peter Asmussen, Denmark
Regional Director:	Barbara Harley, USA
Regional Director:	Slava Khodko, Russia
Regional Director:	Stefan Vrátny, Slovakia

Defense Related SME's: Analysis and Description of Current Conditions
F.D. Carvalho (Ed.)
IOS Press, 2004

The Effect of Military Downsizing on SME's in the Republic of Belarus

Gennadi GAIKO
Republic of Belarus

Abstract. Belarus is a European country located at the border between Central and Eastern Europe. Its population is about 10 million; the capital of Belarus is Minsk with about 2 million citizens. It is one of three Western Republics of the former Soviet Union. Most probably this was a reason for the very high density of military units in previous years.

The military doctrine of Belarus is purely of a defensive nature. At the moment the armed forces in Belarus number up to 83,000 servicemen. The downsizing of 30,000 is scheduled for a five year period.

The number of small companies per one thousand of population in Belarus is one of the lowest in Europe. Although the State programme for small business support and development is annually adapted in Belarus, it is still very difficult for a new business or an individual entrepreneur to enter the market, as well as for an existing company to survive there. Nevertheless, becoming self-employed is frequently considered to be the most effective way to make a living for a person leaving the army.

Military people constitute the most educated and active part of the local population. Military specialization is not always applicable to the new civilian environment. That is why it is not so easy to start a new business. The system of business training, especially for those who look for a new application of personal strength, is helpful.

The BSBI is a non-commercial association of legal entities. The mission of the BSBI is to assist in the creation and development of regional organizations providing support to small businesses, as well as to concentrate, accumulate and complement resources of national and international organizations providing support to business development in Belarus.

The Belarusian Center for retraining and assisting in employment to ex-military servicemen was set up under the Agreement between the Governments of the Republic of Belarus and the Federal Republic of Germany in 1992. The headquarters of the Center is located near Minsk and its regional office operates in the city of Lida in the western part of the country.

There were several attempts to create an NGO with the purpose of representing the interests of those who left the army in Belarus. Unfortunately, none of them were successful.

The experience gained in the countries bordering Belarus in the sphere of social adaptation is very valuable for the country. A practical step the BSBI Association, in cooperation with the Belarusian Center for retraining and assisting in employment, is planning to undertake is the development of a system of entrepreneurial training for ex-military servicemen. The facility of the Center and international links of the BSBI can be effectively used for application of international experience in this field. Donor and sponsor involvement is needed and very welcome.

1. General information

The Republic of Belarus occupies an advantageous geopolitical position in the centre of Europe. In the east it borders on the Russian Federation, in the west on Poland, in the north on Lithuania and Latvia, and in the south on Ukraine. Its territory is 207,600 sq. km. The length of its territory is 560 km from north to south and 650 km from west to east. The length of its borders is 2,969 km.

A greater part of the territory of Belarus is represented by lowland, whereas a quarter of its territory is comprised of hills. Belarus has favourable conditions for human activity: the temperature is continental climate with mild winters and warm summers.

As at 1 January 2001, the population of the Republic of Belarus constituted about 10,019,500 people with more than 100 ethnic groups. According to the 1999 census of the population the Belarusians constituted 81.2%, Russians 11.4%, Poles 3.9%, Ukrainians 2.4%, Jews 0.3% and other nationalities 0.8% of the entire population of the Republic.
The density of population in Belarus is comparatively low, with 49 persons per square kilometer. The rural population constitutes 30.5%.

The Republic of Belarus consists of 6 provinces called Oblast, 118 regions, 102 towns and 108 urbanized settlements. Minsk is the capital of the state with a population of about 2 million people.

The Belarusian language is the language of the indigenous population of the Republic of Belarus. In the 14th–17th centuries the language of the Belarusians was the national language of the Great Principality of Lithuania. After the territory of Belarus was included in the Russian Empire, the Russian language began to spread in Belarus.

For the purpose of preservation of the Belarusian language the Law "On Languages in the Republic of Belarus" was passed in 1990, which proclaimed the Belarusian language the state language. The Russian language was given a status equal to the status of the Belarusian language as a result of the national referendum held on 14 May 1995. Every nationality living in the territory of Belarus is guaranteed the right to study its native tongue and use it.

The Republic of Belarus is a unitary, democratic, social, law-governed state. The Constitution approved the principle of division of powers into legislative, executive and judicial powers. The supreme representative and legislative body is the Parliament. The Parliament consists of two chambers: the Chamber of Representatives and the Council of the Republic.

Over 2,000 non-governmental organizations are registered in the Republic of Belarus. They include 18 parties, 39 republican trade unions and 16 enterprise trade unions, 18 association unions, etc.

Belarus entered the 21st century as a sovereign and independent state, as an integral part of both the European and the world economy and culture.

2. Military downsizing

The military doctrine of Belarus is purely of a defensive nature. At the moment the armed forces in Belarus number up to 83,000 servicemen. The downsizing of 30,000 is scheduled for a five year period. These people will enter the labour market, which is becoming more and more competitive. Although military people constitute the most educated and active part of the local population, military specialization is not always applicable to the new civilian environment. That is why it is not so easy to start a new business.

Some people leaving the army will find a new job easily because they have a good background; they are comparatively young and result-oriented. But many of them will face difficulties in finding a job. Why? If we take into account hidden unemployment, one can say that one in ten Belarusians is unemployed. In these circumstances private enterprises employ one third of people of working age and the number of people going into private business will increase in Belarus in the next few years. But working as a private entrepreneur requires special knowledge and qualifications. The system of business training, especially for those who look for a new application of personnel strength, will work in the country.

3. SME's in Belarus

As at 1 July, 2001 there were 28,300 small enterprises and 153,800 individual entrepreneurs registered in Belarus. The country has one of the lowest ratios of small enterprises per 1,000 population (2.8; compared to 6 in Russia, 46 in England, 68 in Italy, 74.2 in the USA).

The small and medium-sized business sector's role in creating new jobs and improving welfare will only be effective if a comprehensive small business support system is created. It should promote small business development and reduce the number of socially and regionally significant enterprises going bankrupt. The necessary organizational and technical environment, which includes various institutions intended to organize business contacts and commercial operations, currently exists only in a fragmentary form in Belarus.

Since 1994, fifty business development centres, seven business incubators and two technology parks have been registered in the country. But, for various reasons, first of all the lack of international experience and improper understanding of their mission at the local level, they are not fully performing their original functions.

The number of small businesses in the Republic of Belarus has been steadily growing in the last few years. Small companies successfully created jobs: the number of people working in small business increased by 165% from 1997 to 2000. People moved to the small business sector more willingly as salaries were considerably higher than those in State-run organizations, there was a flexible demand for labour and jobs were constantly available.

But there are still a lot of obstacles in the way of SME development in Belarus. Although the State programme for small business support and development is annually adapted in Belarus, it is still very difficult for a new business or an individual entrepreneur to enter the market, as well as for an existing company to survive there. The main hurdles are as follows:

1. Underestimation by government bodies of the role of SME's.
2. Prevailing of fixed administrative methods in relations with SME's and constant spreading of administrative hurdles for entrepreneurial activity.
3. Numerous state inspection bodies, duplication of their functions, extremely frequent inspection of SME operations.
4. Very high level of taxation and very complicated system of state reporting.
5. Lack of reliable legal provisions for SME operations, including proper judicial practice.
6. The majority of the population is not prepared to consider SME's as a tool to overcome social and financial difficulties.

Nevertheless, becoming self-employed is frequently considered to be the most effective way to make a living for a person leaving the army.

4. The BSBI Association

The Belarusian Society of Business Incubators, Business Development Centers, Innovation Centers and Technology Parks (hereinafter referred to as the BSBI Association) is a non-commercial association of legal entities. It was registered in June 2000 within the framework of the UNDP SME Development Project executed in Belarus.

The BSBI`s mission is to assist in creation and development of the regional organizations providing support to small businesses, as well as to concentrate, accumulate and complement resources of the national and international organizations providing support to business development in Belarus.

The BSBI Association seeks to promote business incubators, business development centres, innovation centres and technology parks, to inform and educate others about the benefits of these organizations, to provide information on business development, to monitor legislation, and to bring together key individuals in the small business development "industry".

There were ten founders of the BSBI Association. It is very important to stress that our Association was created with the will of an organization which could deal with problems of small businesses and not with a request from above as happens sometimes. The BSBI membership is permanently growing, and we believe that sooner or later most Belarusian organizations interested in business development in our country will join us.

Our organization already has some experience in implementing projects financed by international organizations. With the first grant we reached a three-fold purpose:

1. conducting a series of training seminars for the BSBI Members in Belarus; the invited speakers delivered their lectures on the development of business incubators, business development centres and credit unions in Central and Eastern European countries (CEE Countries);
2. purchasing office equipment for the BSBI Association;
3. drafting amendments to the Belarusian legislation related to small business support and development infrastructure.

Recently we completed two more projects, which were jointly executed and financed by international donor organizations:

1. .***«CEE countries' Experience in Small Business Support and Development: Workshop for Belarusian Business Incubators and Business Development Centers at Bila Tzerkva Innovation Business Incubator in Ukraine"*** project was implemented in cooperation with the Ukrainian Business Incubators and Innovation Centers Association and financed by the EURASIA Foundation.
2. ***«German-Belarus Business-to-Business Meeting»*** project was implemented in cooperation with the German Association of Business Incubators and Technology Centers **ADT** and Steinbeis Transfer Center for East–West Cooperation, and financed by the German Ministry of Economy and Technologies.

In accordance with BSBI Development Strategy the target-group clubs are being created by the Association. Some of the clubs are of a gender-oriented nature. By the end of last year a women's club was set up to deal with problems faced by women trying to enter small business society. The BSBI Association conducts training seminars, assists the women's club members in compiling proposals for their projects and assists them in making direct contacts with national and international women's organizations.

The next club to be registered at the BSBI Association is an ex-military servicemen's club. It is in our plans to draw experience accumulated in the countries bordering Belarus in the sphere of social adaptation, including financial and consulting resources of our partners in Belarus and abroad.

5. The Center

The Belarusian Center for retraining and assisting in employment to ex-military servicemen (hereinafter referred to as the Center) was set up under the Agreement between the Government of the Republic of Belarus and the Federal Republic of Germany in 1992. The withdrawal of military troops from the Western Group caused the necessity of downsizing. The purpose of creation of the Center was to provide social protection to its target group, i.e., servicemen leaving the army. Actually, the Center has been and is a retraining institution, the necessity for which was mentioned earlier.

The headquarters of the Center is located in the settlement Kolodishchi near Minsk and its regional office operates in the city of Lida in the western part of the country. Before this year the Center was financed by the Federal Republic of Germany, the United States of America and the Republic of Belarus. The international Agreements are now over and the management of the Center is looking for new financial schemes.

To guarantee income for the Center to be able to operate properly, the profit-making educational department was established. At this Department a post-graduate education centre operates. The following subjects are studied at the Department: management, marketing, commercial activity, foreign economic relations, finance and credit, accounting/analysis/audit. The Center also teaches specialists to repair PCs, to repair cars and to process wood. The ex-military servicemen are taught free of charge, and the employment agencies pay for the education of those unemployed people who are enlisted at these agencies.

The Center and the BSBI Association have agreed to work together on projects of further development of social adaptation institutions in Belarus. The offices of both organizations are located very close to each other in the settlement of Kolodishchi, where a lot of ex-military people live. One of the BSBI Members (MAP business incubator) is approaching a final stage of construction of the business incubator building in Kolodishchi, where new companies set up by those who have a new market-related education can be allocated as the business incubator tenants. Another example of cooperation between the Center and the BSBI Association exists in Lida, a town in the western part of Belarus. Three small companies created by ex-military servicemen are located in the local business incubator, which is also a BSBI Member.

6. NGO's in Belarus

There were several attempts to create an NGO with the purpose of representing the interests of those who left the army in Belarus. Unfortunately, none of them happened to be successful. The reason for the failure is simple: it is hard to register and find resources to cover office and management expenses. If the legal entity was registered, the managers would start thinking about the cash flow. If they find a way to commercially make money, they usually forget about the initial idea of their organization. That is why we at the BSBI decided to set up another club for ex-military servicemen.

The BSBI Association keeps friendly relations with those organizations in Eastern Europe which have grown from social adaptation projects, especially in Russia and Ukraine. Additionally, we keep relations with national and international institutions (ministries, UNDP, IFSA) interested in compiling and implementing social adaptation projects in European countries.

The experience gained in the countries bordering Belarus in the sphere of social adaptation is very valuable for the country. A practical step the BSBI Association, in cooperation with the Belarusian Center for retraining and assisting in employment, is planning to undertake is the development of a system of entrepreneurial training for ex-military servicemen. The facility of the Center and international links of the BSBI can be effectively used for application of international experience in this field. Donor and sponsor involvement is needed and very welcome.

Emerging Role of SME's in Russian R&D Conversion

Natalia KOVALEVA
Centre for Science Research and Statistics (CSRS)
Moscow, Russia

Abstract. There are contradictory opinions on the process of conversion in Russia. Administrators of the R&D sector of the so-called military industrial complex consider that its idea and direction are correct. At the same time, managers of enterprises are not satisfied with either the situation today or future prospects. Many conversion problems are related to finance, equipment, personnel, etc. And they are conditioned largely by the poor implementation of the potential of SME's in this sector.

In all advanced countries governments provide small businesses with strong economic and social support. The necessity for such support is recognised in contemporary Russia as well. Over the last 10 years the legal base for SME's has been created. However, the gap between the formal declaration of economic and social responsibility of the State and the real situation is very wide. Data for five years (a regular statistical survey of small enterprises has been conducted in Russia since 1996) show dramatic trends in SME development, and in particular in industry and the R&D sector which accumulates the main conversion activities.

The necessity for market-oriented behaviour makes SME's focus more attention on such managerial problems as strategic planning, analysis and identification of hidden problems, analysis and assessment of the environment, planning of current activities, etc. And there is a strong correlation between managerial problems and the activity of enterprise: the better the results of an enterprise's activities, the higher is the recognition of strong and modern (market-oriented) management.

The hot issues in the market of management training services that require immediate attention and action are: low revenue demand, identified needs for change, out-of-date training and loss of modern training skills, inappropriate structures, irrelevant training, impractical theoretical training, lack of specialised training for intermediaries, necessity of training of teams, etc., and poor information infrastructure.

Introduction

Definition: A small business is a commercial organisation in which the authorised capital includes a contribution not exceeding 25 per cent from the Russian Federation, subjects of the Russian Federation, public and religious organisations (associations), charitable or other foundations. The share owned legally by one or more persons who are not in small business does not exceed 25 per cent and the average number of personnel in the year under consideration does not exceed: 100 persons in industry, construction and transport; 60 persons in agriculture and the R&D sphere; 50 persons in wholesale trade; 30 persons in the retail trade and consumer services; 50 persons in other branches or fields of activity. (The Federal Law "On the state support of small business in the Russian Federation" (1995)).

1. Main trends

The SME formation process in Russia [1] can be divided into several stages. In the late 1970's, some legal opportunities for the so-called "alternative" economic activity appeared outside the State. Ten years later, about 200,000 persons had joined such a kind of business activity. In 1988 the State Law "On co-operation in the USSR" legalised SME's. In 1990 a new impetus was given to the development of small businesses. The governmental decree regarding special measures on the establishment and development of SME's was adopted; one of its main features was the introduction of privileges for small enterprises.

Although all those efforts to reform the Russian economy did provide some special opportunities for SME's, they really did not play any significant role in the Russian economy. 1992 can really only be considered as the starting point for the most recent stage of SME formation in Russia. This period is associated with the beginning of the active privatisation processes in the field of services and trade. The number of such enterprises has rapidly increased, but the process was slowed down by the lack of money in the hands of employers, the lack of credits, the reduction of revenue demand from the population and enterprise, and also by the destructive effects of State policy with regard to taxes and prices. In 1995 the Federal Law "On the State support of small business in the Russian Federation" was issued. By this time the economy of small enterprises had assumed a certain importance, giving rise to the need for a reliable overview of their activity. Since 1996, a regular statistical survey of the activity of small enterprises has been conducted in Russia and its results show the picture that, as a whole, cannot be considered as optimistic [2].

According to the official statistics the SME's production output of goods and market services in 2000 constituted 5.9 per cent (and it has reduced in comparison with 8.0 per cent in 1997) and the investment in the fixed capital – 2.6 per cent (1997 – 5.4 per cent).

Table 1. SMEs by branches (by the end of the year)

	Number of enterprises, *thousands*		Average personnel, *thousands*	
	1996	2000	1996	2000
Total	841.7	879.3	8621.2	7621.1
Manufacturing	**131.9**	**134.2**	**1878.6**	**1621.9**
Agriculture	10.9	14.4	133.2	181.6
Construction	138.0	126.8	2272.5	1586.2
Transport	17.5	19.5	232.9	229.4
Communication	2.9	3.7	30.9	45.0
Trade and public catering	359.3	407.5	2357.5	2607.3
Wholesale trade of products for production & technical use	14.6	13.9	131.7	108.9
Information & computation services	6 1	5.6	46.7	46.9
Real estate operations	3.9	11.1	28.1	81.2
Commercial activities providing operation of the market	35.9	35.7	313.6	218.3
Housing	2.8	5.1	33.2	83.3
Consumer services	10.2	9.6	101.2	92.6
Health care	11.0	18.5	104.4	128.0
Education	6.6	5.5	89.1	47.2
Culture and arts	6.5	8.4	58.2	60.4
R&D	**46.7**	**30.9**	**474.3**	**191.9**
Finance, credit, insurance etc.	10.8	5.9	92.5	33.3
Other activities	26.1	23.0	242.6	257.7

Over the period from 1996 to 2000 the number of small enterprises increased slightly, from 842,000 to 879,000 (see Table 1). But it should be mentioned that the rising tendency was broken in 2000. SME's engaged in one way or another in the conversion process constitute a part of such branches as Manufacturing (46 per cent of the industrial SME's belong to mechanical engineering and metal-working) and Research & Development. The number of the former showed a slow growth trend, while the latter has declined. This resulted in the reduction of share of Manufacturing (from 15.7 to 15.3 per cent) and R&D (from 5.5 to 3.5 per cent) in the total number of SME's.

The majority of small enterprises (93 per cent) are private property and only 6.6 per cent represent a mixed form of property. It exceeds 91 per cent in Manufacturing and 95 per cent in R&D.

A high degree of regional differentiation can be observed in SME networks. By the beginning of 1999 about one-third of the total number of small enterprises (31 per cent) were located in the Central Region (about two-thirds of them being in Moscow) and 14.5 per cent were in the North-Western region (most of those in St. Petersburg). Other centres for small businesses are the Volga, North Caucasian, Ural and West Siberian regions (the corresponding indicator for each of those four regions is nearly 10 per cent).

Small business provides employment in one form or another for more than 7 million people. The total number decreased from 8,639,000 in 1997 to 7,401,000 in 2000. This constitutes only about 12-13 per cent of all employed persons in Russia.

Table 2. SME's Production of Goods and Services

	1997 *mlrd. rubles*	1998 *mln. rubles*	1999 *mln .rubles*	2000 *mln. rubles*
Total	303056.3	261908.0	423711.6	613651.4
Manufacturing	**67895.4**	**63084.8**	**130287.3**	**168098.2**
Agriculture	2245.9	2657.8	4655.4	8167.1
Construction	67579.3	67275.0	96872.8	142480.5
Transport	10529.4	8952.0	13998.0	21161.7
Communication	3035.1	1830.8	5100.2	6575.3
Trade and public catering	102487.4	71113.5	107426.1	158701.0
Wholesale trade of products for production & technical use	7984.2	5219.9	7157.5	10350.2
Information & computation services	1353.7	1273.8	2000.9	5121.1
Real estate operations	1272.4	1604.4	3929.8	7082.1
Commercial activities providing operation of the market	12389.2	7707.5	13692.0	18244.0
Housing	1038.8	1419.2	1737.2	2106.9
Consumer services	1075.4	1559.7	1480.3	2892.5
Health care	3060.7	4954.3	4049.4	7405.4
Education	1340.0	1279.8	1965.0	1604.4
Culture and arts	977.5	1012.0	1424.3	3153.5
R&D	**9796.8**	**9202.0**	**11044.5**	**21819.9**
Finance, credit, insurance etc.	1549.8	5890.6	7215.4	11524.4
Other activities	7445.3	5870.9	9675.5	17163.2

With respect to the structure of small business employment by branches, trade and public catering, manufacturing and construction gained the top positions in 1996. The share of all other branches was 3 per cent or less. 1998 brought changes in SME employment. The most serious decrease occurred in such branches as housing, R&D and transport, while only two branches – commercial activity providing operation of the market and trade and public catering – showed positive dynamics. Over the following years the 1998 shock was not fully overcome.

The SME's total production of goods and services constituted 613651 million rubles in 2000 (see Table 2). And the share of manufacturing in the total SME production grew from 22 to 27 per cent while that of R&D enterprises remained nearly the same and did not exceed even 4 per cent.

In 2000 the SME fixed capital investment reached 29774 million rubles (see Table 3). During this period the share of investment expenditures of industrial SME's grew from 25 to 39 per cent, while that of R&D enterprises comprised no more than 1.5 per cent.

Table 3. SME Fixed Capital Investment

	1997 *mlrd.rubles*	1998 *mln.rubles*	1999 *mln.rubles*	2000 *mln.rubles*
Total	22103.0	19270.0	17887.0	29774.3
Manufacturing	**5668.7**	**3669.7**	**5574.6**	**11592.5**
Agriculture	224.7	208.5	192.8	255.2
Construction	9038.0	6132.4	6129.9	7108.8
Transport	546.7	244.2	313.1	560.1
Communication	528.7	259.9	316.2	379.0
Trade and public catering	3346.0	5608.8	3212.1	4978.0
Wholesale trade of products for production & technical use	475.7	244.0	388.3	267.5
Information & computation services	51.4	23.1	130.8	212.4
Real estate operations	329.5	357.4	364.1	768.4
Commercial activities providing operation of the market	422.0	732.0	167.5	295.7
Housing	158.9	388.7	119.4	62.6
Consumer services	31.3	26.5	34.4	26.7
Health care	201.5	139.7	50.0	2043.8
Education	41.3	44.5	20.1	10.0
Culture and arts	73.0	76.7	33.0	34.7
R&D	**242.4**	**688.0**	**126.2**	**407.4**
Finance, credit, insurance etc.	366.2	275.7	2777.8	375.9
Other activities	357.0	150.4	436.7	395.6

As quite often happens in Russia, the gap between the formal declaration of economic and social responsibility of the State for SME development and the real situation is very wide. Owing to the lack of necessary financial support, as well as the lack of efficient co-ordination between the action of different executors of concrete measures, plus mistakes in organisational issues, the evolution of SME's is not as active as it should be for a real influence on the Russian economy. The position of conversion SME's is more or less the same as other SME's in corresponding branches; they are under the same limitations and meet the same obstacles.

The low innovative activity of SME's is one of the consequences. Only about 10 per cent of Russian enterprises are engaged in innovative activity [3]. It was supposed that SME's (and in particular *Manufacturing* and *R&D SME's*) would be the leaders in the innovation process. But as a rule, the greatest capacity for innovation was shown by large enterprises with a relatively stable financial position and diversified economic activity. Nearly half of innovating enterprises had more than 1,000 employees and 40 per cent had a workforce between 200 and 1,000. In so far as small enterprises were taken into account, the share of small innovative enterprises was considerably lower.

The optimal effectiveness of small enterprises is determined by innovative technological activity, which is not linked to the solution of fundamental problems and does not depend on complex co-operation. The main factors that favour the increase of innovative activity are the following: qualified personnel, strong motivation to engage in innovation activity, good understanding of the market situation and the ability to deal with change. At the same time, poor management, tax burdens, lack of efficient business services for small enterprises, and particularly for SME's in the R&D sector, hamper innovative activity.

2. Conversion: the managerial viewpoint

Conversion issues in the R&D sector are of great importance in Russia. The results of the survey conducted by the Centre for Science Research and Statistics (CSRS) [4] reveal that the managerial personnel – top executives of radio-electronics, aircraft, space and communication industries, as well as managers of corresponding R&D enterprises, expressed contradictory opinions in this respect. The former were full of optimism, whereas the responses of the latter looked rather pessimistic: *the R&D sector "does not play a serious role in this process"; "conversion should be organised in a different way"*. Very often the reforms had a negative effect on the activities of corresponding branches and enterprises: the main components of R&D potential reduced, funding from different sources, the material and technical basis, the quality of personnel and, as a result, the level of their R&D on the whole.

Enterprise managers agreed that the reforms provided some new opportunities for their enterprises; they can independently choose the directions of research, establish links with domestic and foreign partners, use funds, etc. But at the same time the shortage of funds and low market experience did not allow those opportunities to be really fruitful. However, there are examples of positive experience. One of them is the broadening of co-operation with the civilian sector of domestic economy. This was the main "plus", especially for space and aircraft enterprises. Another one is the high interest of foreign countries (potential investors or consumers) in Russian R&D results. The third one lies in the development of the so-called additional market-oriented activities of enterprises, although such activities very often lie far away from the main direction. All those changes called for the implementation of organisational innovation in Russian industry. One of these is the development of small enterprises as an instrument for quick implementation of new products and technologies involving market mechanisms. Different reasons reduce expectations. And regardless of their managerial position, the majority of the experts interviewed could not forecast a good future. To overcome the current situation they suggested changing the structure of activities, paying much more attention to civilian and double-purpose research and production. This would allow enterprises to be more active on the market. And this, in particular, makes the developments of SME's in industry and the R&D sector rather attractive. At the same time, this causes new problems; the lack of strong market experience and corresponding managerial skills restrict the activity of SME's. This problem is common. However, for conversion enterprises it is doubly important because of their previous very strong dependence on State administration and control.

3. Training of Managers for the R&D sphere

Managerial training, particularly the training of managers for R&D enterprises, is of special interest at CSRS. Specific demand and supply issues in the market of corresponding services were the focus of a survey conducted by CSRS. Its results allow some conclusions to be made which can clarify the situation [5].

Russia already possesses many of the prerequisites for a strong innovative climate. The stock of S&T reserves, qualified S&T manpower, incremental reformation and creation of a new legal and standard basis that is currently underway, and the gradual strengthening of government innovation policies, will all help Russia to build a contemporary innovation system.

At the same time, enterprise managers and representatives of educational institutions have made a negative assessment of the innovation climate and point at the main factors that are holding back innovation in Russia. The first of these is the low demand for innovative technologies and products from customers. This is caused both by the lack of enterprises with the necessary current assets and by the fact that many of those involved in the creation of products and services have a poor understanding that a dynamic introduction of innovation is needed in order to ensure high competitiveness. Another problem is inadequate attention paid by federal and regional authorities to science, technology and innovation issues. More efforts need to be made to ensure the creation of innovative technologies and products.

Science and technology endowments do not necessarily lead to innovation; other capabilities, systems and processes are also needed. The need to raise the standard of management applies to all levels of S&T administration and policy-making activity – both federal and regional - and to all levels of activity in particular enterprises.

At present, although it is in principle possible to obtain training in S&T and innovation management, the market for such services needs further development. Such education and training programmes are offered by higher education institutions, in an engineering and economic profile in the context of higher education, postgraduate and additional training. But in fact it is still too early to speak about a well-shaped market of training services.

The much-needed development of managerial training services for S&T and innovation is hampered by a number of factors. These include: the general low level of innovative activity, the poor understanding of the need to improve innovation capabilities on the part of S&T enterprise managers, the low revenue demand from enterprises for respective training services, the inadequate practical relevance of many of the current programmes, and the lack of qualified teaching staff. Furthermore, enterprises display little interest in the employment of young managers who have obtained a state-of-the-art higher education, preferring professionals with practical, although out-of-date experience. One more factor is the low interest of management graduates in S&T enterprises that cannot offer them appropriate remuneration.

Disseminating the results obtained by educational establishments showing real success should support an increase in the level of training. Special attention should be paid to the new managerial techniques and to successful examples of their introduction in Russian S&T enterprises rather than to theoretical studies.

Managerial training needs detailed marketing. This will enable those involved in these services to keep track of, and to respond to, changes in demand for education, including the form and methods used by these services.

The demand from most S&T enterprises for managerial training is underdeveloped. Therefore it is necessary not just to provide services, but to raise at the same time the level of demand for training services. The aim should be to direct potential customers to the acquisition of profound theoretical and practical knowledge which is necessary for current conditions and the prospect of future technological and economic development in general.

It is advisable that the experience of management abroad should be learned as widely as possible. Foreign specialists and teachers should be involved in educational and methodological activities of educational establishments and specialised centres. Their participation in training and upgrading of skills of Russian teaching staff is also desirable.

When implementing foreign training programmes for S&T and innovation management, it is necessary to carry out a pilot programme in Russia. The results of such trials should ensure that the training mechanisms are adapted to the needs of Russian professionals in science, technology and innovation management and education.

It is advisable to give systems of vocational training for managerial staff a multi-stage and multifunctional structure. This will help develop demand from particular customers for training services - as courses can be tailored to suit learners' professional competencies and responsibilities. This will ensure a steady rise in the level of implemented training programmes for S&T management.

It is necessary to improve access to educational and methodological literature on S&T management issues, and to ensure that a sufficient number of copies of the best tutorials are available and recommended for use, through self-education, both at educational establishments and for managers of S&T enterprises.

R&D enterprises are not well informed of the training and skills' development opportunities that exist in management. An interest has been demonstrated in databases that provide easy access to reliable information on the activities of various educational establishments, the implementation of specialised training programmes, training sessions, conferences, symposia and workshops on issues of management of S&T enterprises and the intensification of innovative activities.

References

[1] *Kovaleva, N.*, Small Businesses in Russia. In Small firms and entrepreneurship in Central and Eastern Europe: a socio-economic perspective / Editors Oliver Pfirrmann and Gunter H.Walter. – Heidelberg; New York: Physica-Verl., 2002.

[2] SME's in Russia / Statistical book. Moscow, Goscomstat, 2001 (in Russian).

[3] *Gokhberg, L., Kuznetsova, I.*, Technological innovations in Russia, Moscow. CSRS / 1997 (in Russian).

[4] *Zubova, L., Kovaleva, N., Shuvalova, O.*, Conversion of social and economic issues in the Russian R&D sector / Sociological Research, 1996, 9 (in Russian).

[5] *Gokhberg, L., Kovaleva, N., Kuznetsova, I.*, Innovation Management in Russia. A Review of Training Opportunities for Growth / J. Gristock, E. Bell, L. Gokhberg (eds.) / The British Council, 2000.

System of State and Government Support of SME's in Russia

Andrey A. KRUTKO
Russia

Abstract. The wide practice of utilizing dual-use technologies (DUT) for manufacturing civilian products has stimulated the process of organizational and structural changes of enterprises in the sense of diversification of their activities and development of small business.

Financing of the Federal policy of State Support of Small Business is assured by the Federal Foundation of small business support, which was set up in 1996 by the Federal Government. Its budget is basically made up from the resources of the Federal budget. The Foundation executes orders for drawing up the Federal Programme of State Support of Small Business, including SME's utilizing dual use technologies.

According to the laws of the Russian Federation and of the subjects of the Russian Federation, special privileges exist in taxation of SME's utilizing dual use technologies, in foundations of support of small business, in investment and leasing companies, in credit and insurance companies, as well as in enterprises, institutions and organizations set up for the purpose of execution of works and projects.

Introduction

The wide practice of utilizing dual-use technologies for manufacturing civilian products has stimulated the process of organizational and structural changes of enterprises in the sense of diversification of their activities and development of small business.

The enterprises of the Defence Industrial complex (DIC) traditionally used so-called linear, linear-staff and linear-functional organization of management structures (the terminology used is that accepted in the general theory of management). The inbuilt deficiency of these structures is that they often prove to be very conservative. It is quite difficult to reorient them in the new objectives of the market economy. In addition, their functioning requires additional expense. Introduction of dual use technologies direct DIC enterprises in new types of production organization.

Today many DIC enterprises are launching new structures adapted to market conditions.

Actually, inside the DIC the process of multiplying of SME's oriented in dual use technologies is being accelerated. It is encouraged both by the captains of industry and the Government, which regard it as an effective instrument of guaranteeing employment to the highly qualified DIC personnel, as well as a valuable instrument for making profits for the "mother companies" of these SME's.

According to the Federal Law "On the State Support of Small Business" which came into force in 1996, the subjects of small business are commercial organizations, having nominal capital, the shares of which are possessed by:

a. the Russian State and/or the subject(s) of the Russian Federation and different foundation(s) not exceeding 25%;
b. one or several legal persons who are not subject(s) of small business, not exceeding 25%; the average annual number of employed persons not exceeding 100 persons in industry, and 60 persons in the scientific and technical sphere.

1. Support of dual use technologies

The Russian State has recently been increasing its support to SME's utilizing dual use technologies. In addition to the afore-mentioned Federal Law, the Russian Government initiated adoption of "The Conception of the innovative policy of the Russian Federation" (Russian Government Decree № 832 dated July 24, 1998).

According to the Federal Law, the Russian Federation exercises the right to define general principles, priority directions and methods of State support of small business. The Federal authorities establish the order of setting up and functioning of Federal organs of power responsible for support of SME's, work out and realize Federal programmes of support and development of SME's utilising dual use technologies, which are financed with not less than 50% from the Federal budget and specialized non-budgetary foundations. They also establish privileges for SME's utilising dual use technologies in payment of taxes and other contributions to the Federal budget and specialized foundations of the Russian Federation. The subjects of the Russian Federation (Republics, Regions, Autonomous national districts, the cities of Moscow and Saint-Petersburg) have special rights to introduce additional measures of support of small business and especially of SME's utilising dual use technologies, mobilizing for these purposes their own finance and resources.

The State support of SME's utilising dual use technologies consists of the following measures:

- setting up infrastructure for support and development;
- providing for advantageous conditions of use of the State financial, material and information resources as well as the scientific and technical developments and technologies;
- rendering assistance to the external trade activities of these SME's, including assistance to the development of their trade relations with foreign countries, as well as to the scientific, technical, industrial and information contacts with these countries.
- Federal organs of executive power are responsible for the following actions of assistance to the SME's:
- analysis of the state of affairs in small business and the effectiveness of State support to SME's;
- preparation of the forecasts of SME development and proposals as to how to ensure the amelioration of measures of State assistance to them, and formulation of the corresponding drafts for the consideration of the Government of the Russian Federation;
- organization of the development and realization of "The Federal Programme of State Support of Small Business", provision for the SME's participation in realization of the State programmes and projects, as well as the supply of produce and rendering services for Federal needs.
- As the development of dual use technologies is a State priority, the State assists in R&D works in concrete areas;

- preparation of proposals on the introduction of special privileges, including the sphere of taxation for the SME's and the distribution of budgetary resources and specialized foundation money for their support;
- aid to the executive authorities of the subjects of the Russian Federation in formulating regional programmes of SME support.

State support to SME's is carried out in accordance with the above mentioned "Federal Programme of Support of Small Business", in regional (inter-regional), industry (inter-industry) and municipal programmes of support to small business which are adopted correspondingly by the Government of the Russian Federation and organs of local Government.

2. The Federal Programme of State Support of Small Business

Annually, the Government of the Russian Federation, before introducing draft budget proposals to Parliament, presents to Parliament a draft "Federal Programme of State Support to Small Business".

The Federal budget contains a special provision for allocating necessary financial resources for realization of this Programme "Financial support of the State and municipal programmes of assistance to small business".

This support is exercised on an annual basis with the use of federal budget resources, budgetary resources of the subjects of the Russian Federation, local government and other sources. The volume of obligatory financial resources allocated annually for SME's is mentioned in a separate paragraph in the accounting part of the Federal budget by a presentation of the Government of the Russian Federation.

The Federal Government presents to the State Duma on an annual basis a "Report on the state and development of small business in the Russian Federation and measures of the State support to this business," including "Report on the expenditure of the resources of the Federal budget for State Support of Small Business". There exist and are being created special foundations for assistance of small business.

Financing of the Federal policy of State Support of Small Business is assured by the Federal Foundation of Small Business Support, which was set up in 1996 by the Federal Government. Its budget is basically made up from the resources of the Federal budget. The Foundation executes orders for drawing up the Federal Programme of State Support of Small Business, including SME's utilizing dual use technologies. The Foundation is authorized to issue guarantees on target foreign credits and loans for the support of Russian SME's utilizing dual use technologies, as well as on investment credits allocated by the financial credit organizations of the Russian Federation to SME's utilizing dual use technologies for realization of specific programmes and projects. The profits received as a result of activities of the Federal Foundation of the State Support of Small Business and other State and municipal foundations of this sort are not subject to taxation. The order of drawing financial resources from the State and municipal foundations and issuing guarantees over these resources is established by the foundations themselves under condition of presentation by the SME's of technical-economical and feasibility explication reports of the programmes, projects and measures.

The State and municipal foundations of support to small business possess the right to allocate to the SME's soft loans, interest-free loans and short-term loans without purchase of a licence for bank activities. The basic tasks of these specialized foundations are:

- granting financial assistance on a repayable and interest-free basis for demonopolization and conversion of production with the purpose of

enhancing competitiveness and supply of the commodities market in accordance with current legislation, playing the role of mortgage payer and guarantor and guarantor for obligations of SME's;
- share participation in setting up enterprises and companies dealing with development of market infrastructure and activities, specialized consulting companies, information systems of support of small business and competitivity enhancement and systems of consumer expertise and certification of goods and services;
- financing of training, retraining and improvement of qualifications and skills of the personnel of SME's, assistance to new economic structures, protection of consumers' rights;
- financing of scientific research, scientific practical conferences, symposia, seminars, including international ones and those connected with the activities of the above-mentioned foundations;
- undertaking necessary measures aimed at attraction and effective use of local and foreign investors, including holding of competitions, contests, auctions, exhibitions, lotteries, as well as operations with valuable papers;
- organization of collection and processing of legal, patent-licence and other information useful for SME's, study of internal market conditions, rendering consulting and organizational methodic aid in the development of programmes and projects;
- exercising, according to existing legal procedures, external economic activities, participation in international programmes and projects in the field of small business.

3. Taxation of SME's

According to the laws of the Russian Federation and of the subjects of Russian Federation, there exist special privileges in taxation of SME's utilizing dual use technologies, foundations of support of small business, investment and leasing companies, credit and insurance companies, as well as enterprises, institutions and organizations set up for the purpose of execution of works and projects.

3.1 Accelerated amortization

SME's utilizing dual use technologies are authorized to apply the practice of accelerated amortization of basic industrial funds with reference to expenses of production costs at a rate twice exceeding the norms fixed for the corresponding types of basic funds. Along with the use of the mechanism of accelerated amortization, the SME's have the right to additionally write off up to 50% of the initial cost of the basic funds with a service life of more than three years, presenting it as amortization deduction. In case of cessation of functioning of the SME before expiration of a one-year term of its activity, the additionally-added amortization sums are recovered by means of the increase of balance profits of the involved SME.

3.2 Preferential crediting of SME's

Crediting of SME's utilising dual use technologies is executed with the application of preferential treatment. The foundations of support of small business take up the task of compensation of the tariff difference to the crediting organisations. These foundations enjoy the right to fully compensate the partly crediting organisations with incomes not received from crediting SME's under preferential conditions.

3.3 SME's insurance

According to the law, regulating small business insurance of DUT SME's is carried out at preferential tariffs. The insurance companies involved in this business receive in their turn corresponding privileges and compensations from foundations supporting small business.

3.4 Production for the need of the State

The Federal Law "On the State Support of Small Business" reserves for the SME's specified quotas of orders for production and delivery of certain types of produce (goods). This practice to a great extent is applied to SME's utilizing dual use technologies. The State agencies are obliged to place with SME's not less than 15% of the general volume of orders for purchase and delivery of produce, goods and services for the needs of the State, which have been defined as priorities by the Federal Government or regional governments. These orders are placed on a tender basis.

3.5 Assistance to external economic activities

Federal and regional governments carry out measures directed at broadening of SME participation in export-import operations and realization of programmes and projects concerning external economic activities, such as international exhibitions and fairs. The extent, order and conditions of compensation of expenses connected with the assistance to the external economic activities are established on the basis of contracts between SME's, and the involved State (municipal) foundation supporting small business.

3.6 Assistance in the information sphere

The Federal Government works out and implements measures directed at creation of the corresponding information infrastructure with the aim of granting SME's access to economic, legal and other information necessary for their effective development and providing for exchange of information between SME's.

3.7 Information services to SME's

Information servicing of SME's is carried out at rates beneficial to them. The companies, institutions and organisations delivering information services enjoy the right to corresponding compensation (full or partial) of their expenses connected with servicing SME's.

3.8 Industrial and technological assistance to SME's

Federal and regional governments, as well as the organs of local government provide for distribution of orders for production and deliveries of specialized equipment and other types of produce for SME's. They also render their assistance in setting up and functioning of specialized retail markets, industrial and goods fairs by way of leasing real estate, buildings, structures, equipment and other kinds of property owned by the State or municipalities.

The Power of Partnership

Ian McFARLANE
United Kingdom

Abstract. The attitude of Primes to Suppliers used to be "Arms Length Commercial " and has changed to "Partnership ", requiring much more inter-dependence and closer long-term relationships.

Primes want their suppliers to bring world class capability, in technology, production, price and support, and will help to achieve this.

Even Primes are "Niche Players" and need good, reliable suppliers to form capable teams.

The Role of Offsets. I believe that the use of offsets to create and maintain the above type of capability is valid, but using them to sustain incompetence is doomed to eventual failure.

The Relationships between Government and Industry in achieving these aims.

Introduction

I have entitled my brief address to you " The Power of Partnership", in keeping with Han Willems' "Power" theme. My intent is to try to set a scene looking through the eyes of Primes and to stimulate some discussion.

1. The Changing Policies of Primes

My own background is in working for British Aerospace or its predecessor companies for over 40 years in Guided Weapons, Military Aircraft and, for my last seven years before retirement in 1999, at Corporate Headquarters in Farnborough. Much of these latter years was spent as part of teams working on Mergers and Acquisitions. I give you this background as it influences what I have to say.

For the last 20 years to my knowledge Primes have been reviewing and developing their "Make-Buy" policies. I think this started with the realisation that the smaller supplier companies were much more agile than we were, without the bureaucracies and complex management system overheads that we had and therefore able to respond more quickly to change. We had developed a vast array of sub-contractors and operated an "Arms-Length" trading policy with them as the Government did with us. This policy, while theoretically effective, meant that any disagreement was referred back to the original contract for resolution, sometimes finishing up in litigation, which was good for none of us. Frequently we found that the contract had inadequately defined our needs or that events had forced changes on us requiring contract changes.

As this situation became more and more apparent, we sought for alternatives, among which was dramatically reducing our supplier base from several thousand to a few hundred. This also required that our reduced number of suppliers themselves took on subcontractors and managed them in turn. It also meant that the internal structures of our company changed with some parts effectively being the Primes and others becoming suppliers. Incidentally, I must say here that we had strictly imposed government rules to obey if one of our own subsidiaries was bidding in competition with outside companies, a problem that I know concerns many of you. Our technical selection process was overseen by the customer and we did not see the bid prices, only the MOD had them and was the final arbiter in the selection of the winner.

Little by little I believe that relationships with our reduced supplier base started to show signs of developing trust to replace the old antagonisms of Arms-length contracting and this in turn led to further changes. Under the extreme commercial pressures of the early 90's the concepts of "Partnerships" began to take root. This involved taking our suppliers into our confidence and revealing to them some of our previously guarded secrets. For example, on one project, they were given sight of what we believed were our real marketing prospects, not just for one bid but for worldwide sales, intelligence most suppliers could not create. This had many effects including making them reconsider all their own investments in plant, people and risk and as a consequence the prices they were prepared to bid at.

Other attitudinal changes were also occurring. In the early days a supplier was reluctant to tell us that he had a problem as that usually resulted in arguments and penalties, so we did not know we had a problem until it was on the critical list. This relationship has largely changed as it has been recognised that it is to our mutual benefit to know about problems early and to try to solve them together.

I believe that these changes in the ways of doing business are now quite commonplace among primes and need to be understood by all levels in the food chain.

2. The Demands on Suppliers

The changes I have described in the policies of primes appear at first sight to make the demands on suppliers less demanding. This is not true!! They are required to be more open with their customers about their abilities and their problems. They are required to be prepared for long-term relationships rather than single bid contracts and this often means spending years, perhaps doing minor work, until a bond of confidence is established. Above all they are required to produce world class performance, in my opinion a much misused term. It does not just mean the best technology, but can mean cost, political or industrial influence or product support. Some of these issues are under the control of the companies themselves or their Trade Associations, but many are directly controlled by governments whose role I shall refer to later.

In my opinion, the most important of these, particularly in the smaller nations who have no primes, are matters relating to export. There are few major projects these days where the procuring nation, or even nations in collaboration, can afford to be without the prospect of exports to offset or recoup their development costs. I believe that this is an area where the smaller companies, particularly in the smaller nations must persuade their governments of their needs.

3. The Role of Governments

Governments are the customers for the type of products we are concerned with here, so they can have a fundamental influence on all the issues that concern us by the terms and conditions they impose on their contracts. I commend to you the recently produced "Code of Conduct" produced jointly by the UK government and the Defence Industries Council, defining the treatment to be provide for SME's under MOD contracts. This is an attempt to recognise the problems they face in today's world and a process for dealing with them. I am sure we can deal further with this in discussion.

Governments have also recognised that they need to establish partnerships with their suppliers exemplified in the UK by their adoption of their "Smart Procurement Policy", which seeks to adopt holistic Systems Engineering principles in procurements, again not just relying on simplistic contract relationships to deliver. This, in principle, tries to establish relationships with their suppliers similar to the ones I described for the primes.

Finally, governments are the controllers of export policies, quite rightly. There have been significant attempts to deal with the problems of Security of Supply and Onward Export in international fora such as "OCCAR" and "Letter of Intent" nations, but, as yet, I am unaware of any great breakthrough. As I stated earlier, I believe this to be an area where SME's must put great emphasis.

4. The Role of NATO

NATO itself procures relatively little, most being procured by the nations and placed at the disposal of NATO in times of crisis. Even for what it does procure, this is mostly done through the use of "Host Nations" who use their own procurement rules. I would simply like to leave you here with the question "Would a NATO Code of Conduct for the treatment of SME's", similar to the one adopted in the UK help your cause? (and how would NATO police it?)

5. The Role of Offsets

I cannot leave this session without a word on what I see as the misuse of offsets, a subject on which I have spoken in a number of nations, not always to popular acclaim! I consider that offsets, if used wisely, are very powerful tools. If they are used to create or maintain a world class capability, as previously defined, or to acquire access to world markets, they are wonderful. If they are merely used to sustain incompetence for a short time they are dangerous and only delaying the inevitable demise of that company. To work well there is a need for national policies, agreed to by both governments and their industries, as I believe is the case in Norway,, very successfully. This is another issue I believe should be on the agenda of SME's and their trade Associations.

Support to Small and Medium Business and Development of Innovation in Lithuania

Dr. Pranas Bernardas MILIUS
Kaunas Business Incubator
Studentu 65, LT-3031 Kaunas, Lithuania
Tel. +370 7 300818, Fax +370 7 451599, e-mail milius@ktc.lt

Abstract. When Lithuania finally regained its independence in 1990 the Lithuanian science community began to restructure its science and education system, adapting it to the needs and conditions of an independent state. There are at the present time 15 universities and 29 state–supported science institutes operating in Lithuania. In seeking membership in NATO, Lithuania actively cooperates with NATO and its member nations, not only in the military sphere, but also in scientific research. Lithuanian participation in the "NATO Partnership for Peace" programme helps Lithuania to implement its security goals, to create favourable conditions for its economic development and to expand its cooperation with military associated industries and in the scientific field. Lithuania actively participates in NATO programmes, although its scientific and industrial potential is not being fully used.

There are many projects conducted by Lithuanian scientists, which can form the basis for international co-operation in the defence field. Some of these new devices and technologies developed during the past decade found their place in the civilian markets too. One of the ways to induce this process is support to SME's. The specific features of SME's are favourable for developing and putting innovative ideas into practice. It is very important to note that SME development owes much to different means of government support. The most important of these is preferential taxes and loans. Other means of support are provision of information, training and consulting.

The creation of an Information Society remains a high priority both for Lithuania and other European countries, but a great lack of IT professionals has been noticed. Therefore we aim to create the most favourable conditions for entrepreneurs to start SME's in the priority field of IT. Obviously, in this case, specialised IT Business Incubators and Science Parks could effectively support a number of innovative SME's.

It is important to note that the majority of Lithuanian SME's have lost their markets in the East, but gained a position in the West. Rapid changes in sales markets foster flexibility and innovativeness of enterprises. However, enterprises lack a strategic innovation implementation model that would combine empirical methods and methods of strategic planning that have as their purpose the promotion of innovation development and analysis of SME expansion perspectives.

Introduction

When Lithuania finally regained its independence in 1990, the restructuring of both the Lithuanian economy and the science and education systems began adapting to the needs and conditions of an independent State. A considerable number of large enterprises stopped activities that had been too dependent on orders from other former Soviet Union countries and could only function in a centralised economy. At the same time. many SME's were started – more or less successfully, but with no government support at the beginning of their activities. It took several years to acknowledge the SME's importance in the local economy and to start to support them. It has also been recognised that SME's are often the best place, not only to create new working places, but also to implement innovation. Internationalisation of business and scientific activities is also starting to bring positive results, inducing further developments, especially in co-operation with EU and NATO countries, and showing a large innovation potential in Lithuania that can and must be used, not only domestically. SME's and innovation support activities in Lithuania are still rather limited due to the lack of financial resources and other reasons; however, development of a support system is progressing.

1. SME's impact on social and economic indicators

The specific features of SME's are favourable for developing and putting innovative ideas into practice. Moreover, the creation of new working places owes much to the SME sector expansion. At the beginning of 2001, SME's made up nearly 94% of Lithuanian operating enterprises (numerically it was about 64,000). The majority of these are to be found in the largest regions, e.g. Vilnius-the capital, Kaunas–the second largest Lithuanian city, the region of technologies, Klaipeda -the seaport, Siauliai and Panevezys.

From the perspective of membership in the European Union, we are more and more concerned about the peripheral regions of Lithuania which face severe problems of unemployment.

It is very important to note that SME development is influenced by different means of government support. The most important of these is preferential taxes and preferential loans. Other kinds of support are the provision of information, training and consulting.

Under the initiative and/or financial help of the Lithuanian Government the following SME support institutions have been established and are functioning (see Figure 1: SME support infrastructure in Lithuania):

- 7 Business Incubators (with significant co-financing by Government);
- 6 Business Advisory Centres (BAC) closely dealing with the Lithuanian SME Development Agency (at the moment many BAC are being reorganised into Regional Development Agencies);
- 3 Business Information Centres (Zarasai, Jurbarkas, Svencionys); 40 new centres are foreseen to be established in the near future;
- 2 Euro-Info Centres (Vilnius, Kaunas).

In addition to premises rented under favourable conditions and the organisation of training and consultations for businessmen, it is extremely important to grant credits, especially at the start of companies' activities.

1.1 Credits for SME's

- 4 million EUR are assigned for preferential loans (more than 400 loans were awarded to the companies from 1993-2000);
- More than 30 Credit Unions are operating in Lithuania at the moment;
- 21 million Litas were granted to SME's by the Lithuanian Government from1998-2001;
- Approximately 70% of local authorities are managing their own SME Development Funds (2.9 million Litas in 1999, and 2.8 million Litas in 2000).

Financial support to SME's is mostly provided by means of a Privatisation Fund. However, a large part of these funds is not used to finance investment projects, but:

- To purchase State Stocks, i.e. to cover a deficit in the national budget;
- To repay the saving of citizens (up to 2/3 of all means);
- To cover a deficit in the Social Insurance Fund;
- To support culture and for renovation of public health institutions.

Money in a Privatisation Fund is split into different programmes, so that only a relatively insignificant part is left for SME support purposes.

Referring to reinvestments made in companies, most of these are used to purchase new equipment, buildings, PCs (hardware & software), and means of transportation.

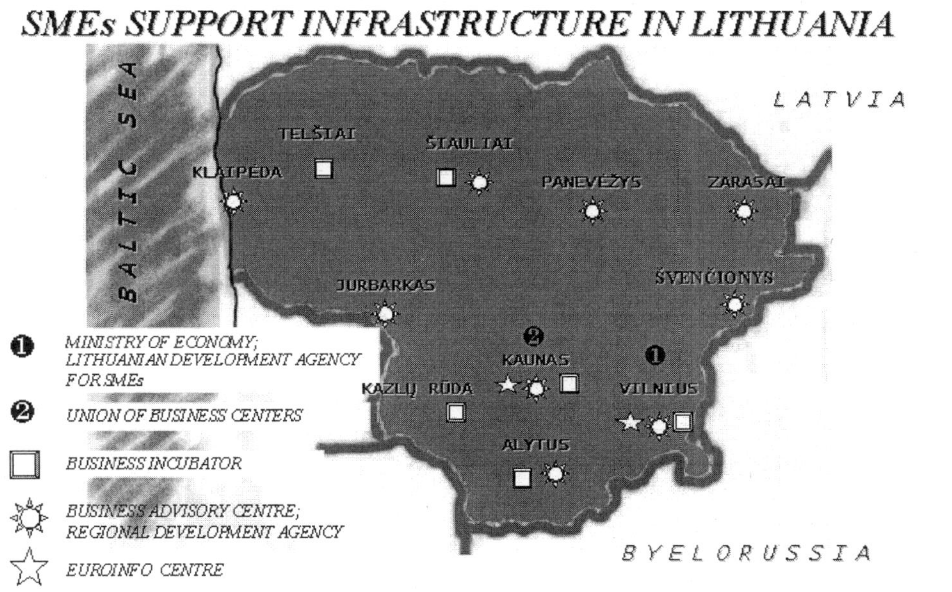

MINISTRY OF ECONOMY	UNION OF BUSINESS CENTRES
GEDIMINO PR. 38/2, VILNIUS, LITHUANIA	STUDENTU ST. 65–302, KAUNAS, LITHUANIA
TEL.: (+370-2)623940, FAX: (+370-2)623974	TEL.: (+370-7)300828; FAX: (+370-7)451599
EMAIL: verslas@po.ekm.lt	EMAIL: sekretor@ktc.lt

Figure 1 - SME support infrastructure in Lithuania

1.2 Creation of an Information Society

The creation of an Information Society remains a high priority both for Lithuania and other European countries; however, a great lack of IT professionals has been noticed. At the same time, the migration of Lithuanian IT specialists has already begun and will be more intensive in the future. Our State is concerned about saving these specialists by creating new working places; economic conditions, however, are not favourable enough to stop the so called "brain drain".

With regard to the means taken for the successful development of an information society, the following can be named:

- A special programme of vocational training for IT specialists should be prepared. This programme should be focused on increasing the number of IT students in colleges and universities which offer Master and Doctoral studies;

- This programme should obtain support at the highest levels of the Lithuanian Parliament and Government. The possibilities of national financing should be explored.

In 2001 a Forum to promote a knowledge society was established in Lithuania, with the aim of creating the best environment for the development of a knowledge economy in Lithuania:

- To improve co-operation between science and knowledge-based business,
- To improve the education system which should correspond to the needs of modern society and modern business,
- To create a transparent and open business environment and free competition,
- To disseminate information on a knowledge economy to society.

1.3 Business Incubators and Science-Technology Parks

As was mentioned above, the lion's share of GDP is created by SME's. The majority of the newest SME's operates in the field of information technologies which is so important to our State with its poor natural resources. Unfortunately, this sector of enterprises does not receive enough attention from national and local authorities.

Therefore we aim to create the most favourable conditions for entrepreneurs to start small and medium-sized companies with priority in the IT field. Obviously, in this case, specialised IT Business Incubators and Science Parks could effectively support a number of innovative SME's. Therefore national and local authorities should be concerned about establishing and strengthening these institutions. It means that a special programme of IT BI and SP foundations, together with a special Fund, should be developed. A fixed part of a Privatisation Fund should be set aside to solve unemployment problems. Legislation concerning BI and SP should also be corrected and renewed.

2. Development of innovation activities

Increasing interest in innovation is closely related to scientific-technological progress, to a greater receptivity of SME's concerning technological novelties, and to highly qualified personnel. Analysis regarding transfer of novelties and putting them into practice, showed that these processes can be very fast due to the SME's' wish to gain a competitive advantage. It is important to note that such behaviour of SME's is related to the economic processes in sales markets; the majority of Lithuanian SME's have lost their markets in the East, but gained a position in the West. Rapid changes in sales markets foster flexibility and innovativeness of enterprises. Enterprises, however, lack a strategic innovation implementation model that would combine empirical methods and methods of strategic planning, the purpose of which is to induce innovation development and analysis of SME expansion perspectives.

Implementation of innovation is closely related to the large rate of business risk. It is important therefore to obtain a State guarantee of credits for new technologies and novelties' implementation in SME's. For this purpose the Lithuanian Government founded a special guarantee institution, the Business Development Company (BDC), which will provide commercial banks with a national guarantee. The total capital of the BDC is 10 million Litas and 8 million make up the authorised capital.

Development of innovation has also much in common with the financial support of science and RTD results that are utilised to accelerate technological, economic, and social progress of the State. It is asserted through national support of Business Incubators, co-financing of other different projects under the "EUREKA", "FP-5" and other programmes. According to the national government programme of 2001-2004, great attention is being paid to innovation development; it foresees that financial support to scientific research activities should be supported on the principle of programmes, stating also the need for integration of science and business, in particular, SME's.

In seeking membership of NATO, Lithuania actively cooperates with NATO and its member nations, not only in the military sphere, but also in scientific research. Lithuanian participation in the "NATO Partnership for Peace" programme helps Lithuania to implement its security goals, to create favourable conditions for its economic development and to expand its cooperation with military associated industries and in the scientific field. Lithuania actively participates in NATO programmes, although its scientific and industrial potential is not being fully used.

There are many projects conducted by Lithuanian scientists which can form the basis for international cooperation in the defence field. Some of these new devices and technologies developed during the past decade found their place also in civilian markets. One of the ways to induce this process is to support innovative SME's.

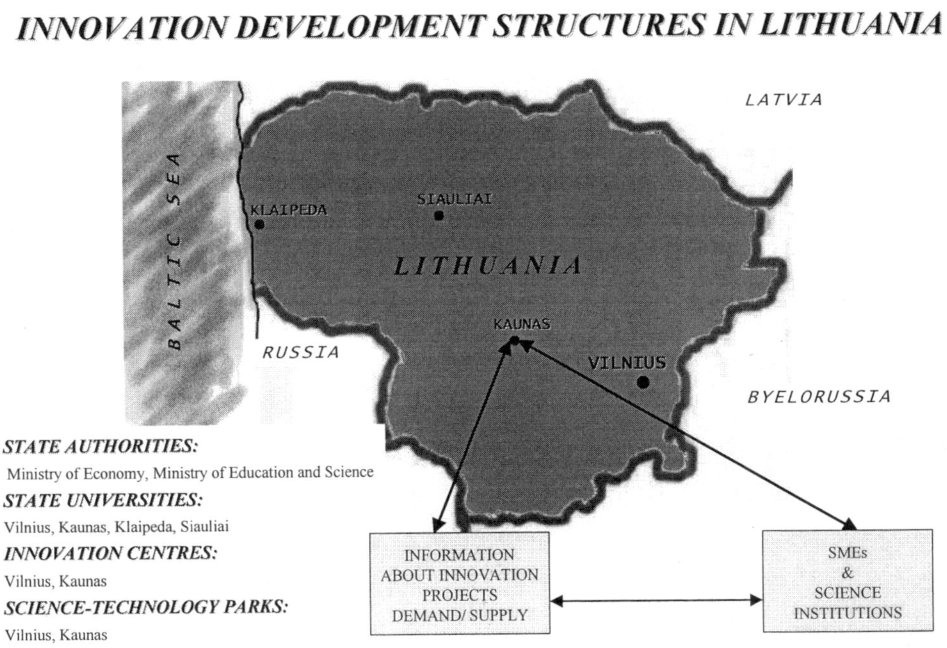

Figure 2 - Innovation support/ development structures in Lithuania

The innovation support system in Lithuania is based on cooperation between several kinds of institutions: State authorities (ministries, municipalities), universities, business incubators and science/technology parks. Innovation support can be divided into political support (programmes, concepts approved by the Government), financial support (3 financial support programmes of the Ministry of Economy), and information, consulting services and infrastructure for innovative SME's (innovation, advisory, Euro-info centres, business incubators, science/technology parks). Despite efforts already undertaken, there is still much to be done in this field. The main problems are lack of cooperation between science and business, a not yet fully developed State innovation support system, a lack of business confidence in innovation, and unavailable venture capital.

For innovation support/ development structures in Lithuania – see Figure 2.

3. Conclusions

The promotion of SME's and innovation development policy should be focused on the following:

- United efforts of authorities, science and business associations and other interested parties have to be used in preparing long-term SME support and innovation development strategy;
- Business and scientific institutions must systematically diagnose and evaluate in favour of economic policy dealing with innovation and business promotion;
- Precise expertise and standardisation of Lithuanian legislation must take place in order to avoid and exclude any contradictions and repetition; it is necessary to ratify the special institution that would be responsible for innovation policy formation;
- Initiate a special programme for new working placė creation in IT and other related fields of activities; for this purpose specialised Business Incubators and Science-Technology Parks should be used;
- When financing research and developing national research programmes in Lithuania, priority must be given to applied research oriented to the creation of new technologies and innovation and their further use for Lithuanian industry needs;
- A legal basis should be created that would stimulate industry enterprises and firms to support applied scientific research, the practical use of its results and technology transfer.

Utilisation of Left-Over Military Sites in Hungary

Péter MOGYORÓSI, Márton Vilmányi, Balázs Révész
Laser Consult Kft. and the Faculty of Economics at the University of Szeged, Hungary

Abstract. Left-over military camps can be grouped as (i) ex-Soviet camps and (ii) Hungarian camps because of downsizing. Utilization of these military camps has different models, such as converting them to

- Industrial parks
- Science parks
- Logistic centres
- University campuses
- Civil airports

For SME and Regional Development purposes the most interesting model is the industrial park.

The first group of problems was the ownership of the property. The ownership was

- given to the local municipality, or
- given to the local university, or
- sold to investors.

The second group of problems concerned environmental issues. Left-over ammunition and fuel contamination of soil were the major problems.

The third group of problems consisted of outdated buildings and other constructions (infrastructure, parade fields, etc.).

After solving all these problems, new investments could be initiated.

Case studies will be presented for different solutions.

1. ex-Soviet military camp converted to a successful industrial park in Vecsés
2. ex-Soviet military camp converted to logistic centre in Dunaföldvár
3. Hungarian military camp converted to University campus in Szeged
4. ex-Soviet airport converted to a civilian airport and a connected industrial park in Tököl

Introduction – The issue of abandoned military sites in Hungary

It is only a distant memory today that, when touring the country, visitors passed by Soviet barracks in Hungary. There were rumours about the prevailing conditions behind the walls, about the punishment of young soldiers, about the one and a half metre high, dark and dank detention rooms (whose floors were an inch deep in water) where the detainees had to crouch, about the Soviet officers' strange barter trade, and the formidable Soviet war machine. The "constant" standby of the Cold War era is now a thing of the past. With the dissolution of the Warsaw Pact, the Soviet army – in which the number of soldiers amounted to the population of a medium-sized town – left the region, leaving behind some 288 military facilities in 104 settlements. The registered value of the facilities slightly exceeded USD 300 million. However, to be converted into actual value, the sites had to be cleaned by bomb-disposal squads and chemical clean-up specialists.

More sites have become available as a result of the reform of the Hungarian armed forces, and their civilian-economic use presents further problems for the reutilization of these barracks.

The most important question now concerning the abandoned military facilities is how to make use of these sites socially, how to reintegrate them into the civilian sphere and make them an organic part of it, how to exploit their available features and turn them to good advantage.

The principal issues of this programme are as follows:

- firstly, who should decide on the utilisation of the sites as the proprietorship of these real assets needs to be carefully looked at.
- secondly, what features these real assets have; this question reviews the potential possibilities and limitations of the utilisation.
- thirdly, which models have determined the use of the sites so far, to what extent these models are justified, and in what measure they contribute to the realisation of set goals.

Our study really seeks to address the above issues. Later, case studies shall be provided as an illustration, in order to present the models employed so far.

1. Property relations of the vacated military real assets

The issue of property relations of the military facilities will be dealt with in two distinct spheres, and this separation shall be maintained throughout:

- firstly, with regard to the former Soviet barracks, where the property rights were once exercised by the Soviet Red Army, the real or potential user has been unclear from the moment of the handing-over.
- secondly, the sites have been sold or handed over by the Hungarian national defence. As a consequence, until the handing-over the property rights have been/are unequivocally exercised by the Ministry of Defence, and during the handing-over they were conferred upon the user of the respective sites.

A Government Decree from 1991 on settling the issue of ownership of the Soviet barracks stipulated that the facilities built by Hungarians prior to the Soviet occupation should become the property of the municipalities, while those built by the Soviets should become the property of the then Treasury Assets Management Organisation. By making this decision, the Government managed the distribution of assets used by the Soviet Army. The real assets taken from the local communities have been returned to them, whereas the "new" infrastructure established during the period of the Soviet presence was brought under central control.

In June 1995 some 150 Soviet barracks, airports and other real assets, which remained in central control, were transferred from the Treasury Assets Directorate to the Hungarian State Privatisation and Holding Company (ÁPV Rt.), which was established for the professional management of state property.

After being taken over, the real assets had a registered value of about USD 85 million, but due to their decrepit state they represented a market value of barely USD 20 million. In 1996 ÁPV Rt. realised several sales of assets by tendering them to a fairly mixed variety of buyers. These sales took into account not only the profit angle, but also the plans for further utilisation.

The inherited real assets were remarkably complex and varied. Moreover, in certain cases, the location itself sometimes also hindered the sale of the sites. The costs of real asset management and maintenance proved to be rather high, however. The only viable solution seemed to be to transfer these facilities into real ownership as quickly as possible.

Taking the above into consideration, ÁPV Rt. initiated an amendment to enable the handing-over of certain former Soviet military sites to the municipalities free of charge. The reason for the handing-over was primarily so that real assets of low market value and high management costs should be transferred to an owner – the municipalities in this case – who would be able to make more effective use of them than the State, and would benefit local communities. Section 8 of the Budget Act of 1997 expanded the range of assets that could be handed over free of charge to the municipalities. The Act categorised the former Soviet barracks according to their intended function. Municipalities could obtain those real assets, which were found suitable for residential development and for other welfare purposes. A contract was concluded prior to the handing-over in which the municipality undertook to use the facilities according to the planned purpose. Another possibility was for the municipalities to claim the assets for other public purposes. In this case they could acquire the ownership by means of an individual government decision. After the said individual government decision was handed down, the former Soviet – non-saleable – properties became transferable free of charge to the municipalities.

Since the handing-over, all responsibility has been assumed by the municipalities with the exception of the addressing of any environmental damage.

In the case of certain facilities, however, the asset reorganisation costs were so high that not even a propriety transfer free of charge could ensure their proper utilisation by the municipalities. Depending on their location, ÁPV Rt. has sold these real assets – after the demolition of the existing structures and the reorganisation of the area – as arable land or as an area of common use.

Certain real assets were placed at the disposal of the municipalities excluding those sold in 1996, those qualified as unsuitable for utilisation due to the lack of reorganisation, and those currently owned by ÁPV Rt. owing to the lack of an accepted utilisation concept.

As regards the former Hungarian barracks or assets of military purpose, this procedure was somewhat simpler. As a result of the national defence reform, the marketable real assets were presented in the budget of the Ministry of Defence as incomes from domestic sales (the sale of these sites was required to keep those national defence expenditures considered necessary within limits). This motivated the Ministry of Defence to ensure a speedy sale and handing-over of the real assets to the end-user. The Ministry of Defence solved the issue of selling these real assets by annual tendering procedures, although it placed restrictions in the form of easement or obligatory utilisation purpose for certain real assets. The end-users of these real assets have turned out to be a fairly mixed group. These sites have been transferred primarily to private investors, institutional investors, companies or municipalities.

2. The features of the vacated military facilities

The military sites have mainly become the responsibility of municipalities (except those handed over to institutional or private investors). The initiation and implementation of utilisation programmes – that exploit the local conditions and possibilities and thus best serve the interests of the local communities – has become the responsibility of the municipalities. These programmes posed a great challenge to institutions that had scarcely any disposable income. When examining the range of available possibilities, they had to perform a tailor-made analysis of the existing features:

- the issue of location
- the issue of infrastructure
- the issue of standards
- the issue of environmental protection

A thorough review of these issues required great foresight, since the question of utilisation is not just a problem affecting the needs of local communities, but also a significant cost-utility problem that has to be addressed.

2.1 The issue of location

The former Soviet (as well as the former Hungarian) barracks are mainly located on the outskirts of settlements or in their environs. Their location often presented an obstacle for a quick sale since the possibility of speculative trading or a value preserving investment was quite limited for these real assets. At the same time – taking into account the general state of the facilities after the handing-over – considerable investments were required to enable any conceivable production activity. Owing to the close proximity of the real assets to towns, the responsible municipalities were under constant political/electoral pressure to find a solution for the utilisation of the increasingly dilapidated facilities. It was also not a viable option to sweep the problem under the carpet by indefinitely postponing decisions about their utilisation. Nonetheless, their location offers some advantages owing to their former military use. These include the close proximity to good motorways, major railways, water and air traffic or traffic zones, and perhaps the availability of an airport or railway loading facility nearby.

2.2 The issue of infrastructure

The issue of infrastructure had many sub-problems: firstly, the constructed facilities themselves (residential buildings, barracks, or other buildings) and, secondly, the installed infrastructure (sewage, water and electricity supplies, telephone network, etc.). The chief issue for both the Hungarian and former Soviet military facilities was their poor condition. When the local utilisation decisions were reached these structures had been empty for at least 5 years, without any kind of maintenance or repairs being done (it should be noted that even the newest buildings were built in the mid-eighties and are on average 20-30 years old). The public utilities necessary for the proper running of the real assets were available (at least at the site boundaries, and this represented a great advantage for later industrial utilisation), but even their condition reflected a very sorry picture. A further problem was that the persons who handed over the facilities removed the equipment (pumps, ventilation, transformers, etc.) necessary for public utilities to function properly. Moreover, the handing-over took place after this equipment had been removed. (There were some notably shocking cases where the withdrawing Soviet army dismantled and transported even the window frames, water-closet pans and other elements constituting an organic part of buildings. One should also mention that in the case of the sale of former Hungarian barracks the dismantling occurred in accordance with the agreement made with the purchaser.)

2.3 The issue of standards

As regards the infrastructure, the disparity between the Soviet and Hungarian building and construction standards represented a serious problem. The investments initiated and carried out by the Soviet military command were done on the basis of the standards valid in the Soviet Union at that time, which did not meet the Hungarian standards even at the time of their inception. (These standards posed structural, roof structure, insulation, etc., problems for buildings. *A propos* other infrastructure problems, these included electric shock protection and environmental protection.) These issues emerged at facilities that had been set up entirely under Soviet supervision, and at other existing facilities taken over by the Soviets.

2.4 The issue of environmental protection

The most serious issue of all was the question of environmental protection – and it still is today. The above-mentioned real assets were seriously damaged because of the improper handling of fuel and explosives over the past few decades. The contamination was mainly brought about by the following compounds: TPH (total petroleum hydrocarbon); PAH (polyaromatic hydrocarbons, without naphthalenes); benzene (according to the measurement results there were cases when the concentration exceeded the set limit by 200 times); ethyl benzene; alkyl benzene; xylene; (results show that it exceeds the prescribed limit by 5-20 times). Among pollutants one should also mention chloroform (which exceeds the set limit by 5-10 times on average) and 1.2 ethylene dichloride (which exceeds the set limit by 5-20 times). The above pollutants had to be dealt with before the beginning of further utilisation of a site (according to Act 33/2000 by the person who owns the particular property). The cost of cleaning up a site is USD 1 million on average, a sum which – with the proper professional preparation and justification – is provided by the Government. It will put contracts out to tender and pay for them from funds earmarked for environmental purposes.

3. The issue of social usability of the abandoned barracks

The sites, viewed as real assets, fell under the municipalities' sphere of authority. After assessing the existing conditions, the local decision-makers had to choose from the possibilities below:

- utilisation of the real assets as a means of direct economic development
- utilisation of the real assets as a means of indirect economic development
- other modes of welfare, social or business utilisation

The motivation for economic-regional development proved to be a priority and a predominant aspect of the utilisation. The reason for this can be traced back not only to the (primarily industrial) features of the said real assets, but also the ever-present political will for economic development which characterised the 1990's and was propelled by the urge to place the disordered economic units onto a path of growth. The primary intention of the organisations that disposed of the sites was to make use of these real assets directly or indirectly for the benefit of economic development purposes (these intentions prevail even today). Utilisation attempts for some other purpose were made only under special circumstances or within the framework of forced, second-best solutions.

3.1 Utilisation of the real assets as a means of direct economic development

By the mid-1990's the Hungarian state-owned, predominantly large-scale enterprises largely disintegrated and their remaining units were either privatised or closed down. Local venture capital, however, was by no means strong enough (and it could not have become so) to fill the gap left behind by the formerly existing large enterprises, or to set a new course for further development in their field of activity.

The lack of available capital, management skills, and new technologies was noticeable throughout the entire country, which required the inflow of working capital. Apart from easing the apparent shortage of capital, the political and economic decision-makers expected from the inflow of working capital the intensification of supplier and co-operational contacts as well as general improvements in the fields of production management, management methods, technology, corporate information systems, product quality, production profiles, competitiveness and preconditions of market penetration. Outside the available privatisation possibilities, plots of land suitable for greenfield investments were to be made available in order to ensure the required effectiveness of the inflow of working capital. These sites provided favourable conditions for such investments owing to their close proximity to traffic junctions and established infrastructure. (The importance of these investments is indicated by the fact that about 30% of the entire working capital flowing into Hungary was invested in the form of greenfield investments.)

A further issue was the need to establish an institutional system focusing on the protection of the "fledgling" domestic capital and its businesses by allowances and value added services.

To guarantee the above conditions, a system of industrial and technology parks was established in the second half of the 1990's as well as the incubators whose range of services was aimed primarily at the latter target group. The available real assets formerly used for military purposes (where the constructed or installed infrastructure was refurbished via government aid) served as an ideal background for establishing such incubators. The industrial park established on the site of the former Soviet barracks in Tököl, on the outskirts of Budapest, provides an excellent example of how these real assets might be utilised as industrial parks. As for technology parks, the incubator and business development centre erected in Nyíregyháza on the site of a former Soviet barracks is also a good example.

The Government played a significant role in the establishment of these facilities and in making them available. By creating the qualification system for the industrial parks, infrastructure investments and environmental recultivation as well as funding programmes for establishing incubators, the Government considered supporting the utilisation of former military sites an issue of overriding importance. (The subsidy provided was USD 0.5-1 million on average, which included the financing of the above-mentioned investment aims.)

3.2 Utilisation of the former military facilities as a means of indirect economic development

The utilisation of the sites as a means of indirect economic development became a reality in the latter part of the 1990's. It became obvious that a significant background of infrastructure would be required to initiate the next phase of development of the Hungarian economy, especially in the fields of:

- the creation of the infrastructure background necessary to ensure the transfer of knowledge and know-how
- the creation of the infrastructure supporting the spatial distribution of the input-output processes of the players in the economy

The latter can be better understood after grasping two key points:

- Firstly, at the beginning of the 1990's it became apparent that on an international comparison scale the ratio of the principal vehicles of the transfer of knowledge and know-how, namely the percentage of those obtaining a university or college degree, was very low in Hungary. To tackle this issue, the Government committed itself to the continuous increase in the number of students in higher education from 10-15% of the characteristic age group, which was typical of the early 1990's, to some 50-60%. This generated enormous capacity problems in Hungarian higher education at the end of the last decade, which had to be urgently addressed. Instead of building new facilities, the university towns now had the rational alternative of reconstructing the existing military real assets – which were furnished with the necessary infrastructure – and thus utilise them for higher education purposes. In this way it became possible to establish university buildings and in some cases dormitories (through the refurbishment of flats of former army officers) at these sites. There are two splendid examples of this: the founding of the campus for the Faculty of Economics at the University of Debrecen and the establishment of a dormitory-complex belonging to the University of Szeged, both at the sites of former Soviet barracks.

- Secondly, by the latter half of the 1990's it also became clear that the knowledge transfer between the corporate, educational and research institutions – which was rather poor according to numerous surveys – would need to be boosted by every means necessary. Otherwise knowledge transfer would be restricted to the companies of foreign interest (i.e. those companies that benefit from the knowledge transfer coming from abroad would be foreign ones). Thus the Hungarian business sphere would have even fewer opportunities to gain insights into the latest scientific and technical innovations. In this respect there was a shortage of the so-called bridging institutes which are able to mediate "knowledge market" services with high added-value to the corporate sphere in a usable way. To tackle this shortage a dynamic government programme was initiated for the establishment of innovation industrial parks and R&D incubators. In addition to the research institutes and the business sphere dealing with R&D, higher education institutes had a vital interest in the realisation of this programme. Owing to the lack of financial resources, the higher education institutes taking part in these initiatives generally placed their "freely available" real assets – i.e. the unused former military sites – at the disposal of those in charge of the project's realisation. A good example of the initiative mentioned above is the innovation park established by the municipality of Szeged and the University of Szeged at the site of a former Soviet barracks as well as the R&D biotechnological incubator, which is currently under construction. Another example is the innovation centre also established at the site of a former Soviet barracks as a result of joint efforts of the College of Dunaújváros and the municipality of Dunaújváros.

Certain former military sites with a special status were made available for utilisation in order to establish a concentrated infrastructure promoting the spatial distribution of the input-output processes of the players in the economy. The development and potential development of road, rail and air transport routes was the primary concern in this case. Where these infrastructure conditions were available, the former military facilities could be turned into logistics service centres or logistics parks. A central location at traffic junctions was a significant aspect in this utilisation model (for example the availability of a river harbour or airport, the proximity of a bridge over a river as well as the proximity of a rail distribution centre). In addition to the active participation of the local and county municipalities, these projects were realised in the form of an industrial park programme primarily from government subsidies already referred to. A fine example of this is the establishment of the logistics centre in Dunaföldvár. The key Danube bridge in the southern part of Transdanubia, the available harbour capacity and the real estate capacity of the former Soviet military site built in close proximity to of all these provided ideal conditions for the establishment of a logistics centre.

3.3 Miscellaneous modes of welfare, social or business utilisation

Among these the following utilisation models should be highlighted:

- Utilisation for housing purposes, the establishment of residential parks:
  Although treated here in the "miscellaneous" category, the utilisation of former military sites for housing purposes can be regarded as the first concept that emerged. A considerable proportion of mostly former Soviet barracks was utilised for the purpose of providing accommodation for the Soviet officers and their families. After the political transition the departing Soviet Army left behind about 15,000 real assets used for residential purposes. Certain real assets were refurbished with the help of investments by local municipalities, commercial banks, or local venture capital (apart from the optional governmental subsidy of 80-85%). In this way the problem of lack of available housing for the younger generation was alleviated. By today approximately 80% of these real assets, which were initially in a rather poor state, have been refurbished by enormous efforts and have been handed over. (It should be remarked that the real assets handed over to the universities represent a part of the real assets originally constructed for housing purposes; about 5-10,000 flats have been utilised for housing purposes). Examples of such utilisation programmes were employed at the former Soviet barracks in Kecskemét, Veszprém, Dombóvár and Debrecen.
- Sale to investors:
  The sale to investors has been the most common way of utilising the military real assets that became available as a result of the reform of the Hungarian armed forces. This sale is certainly a necessity for the Ministry of Defence: firstly because their maintainability has been called into question and, secondly, due to the need to finance the national defence (as the central budget relies on the marketing of these real assets and considers the incomes derived from them as additional resources). During a sale the Ministry prescribes the maintaining and utilising of the available infrastructure by setting a specified usage in the contract (for instance, most airports shall remain airports), and maintains further easement rights as well. The investors are mainly business investors with industrial investment aims (mostly for abandoned barracks) or tourism investment aims (mostly for military holiday resorts and sports facilities intended for sale). The purchase and investment made by Audi in 1999 is a fine instance of this type of investment.
- Temporary leasing:
  This form of utilisation is not typical for real assets, which once had a military purpose, but rather for tourism-related facilities – mostly real assets constructed for holiday, sports and catering purposes – that were transferred to the Ministry of Defence.

4. Case studies

4.1 Utilisation as an industrial park

Due to the features of the facilities once used for military aims, utilisation for economic development purposes is a mode of utilisation of prime importance. Beyond the sale of these sites to economic enterprises, the Government also supports within this framework the establishment of industrial parks by means of transfer or sale of the real assets to the local municipalities. Today there are approximately ten industrial parks in Hungary operated partly or entirely on former military sites.

The **Dunaföldvár Industrial Park** was established at the site of a former Soviet barracks in 1998. The industrial park functions primarily as a logistics centre due to its favourable geographical position. The town of Dunaföldvár is located right in the middle of Hungary, on the banks of the Danube. Its traffic network location is highly favourable as the town is intersected by one of the so-called European corridors, namely the No. 6 main road. Here, south of Budapest, lies the first bridge over the Danube that connects the eastern and western parts of the country.

The Dunaföldvár barracks was originally a real asset belonging to the Hungarian Army. During the period following the Second World War it functioned as a national defence barracks. After the repression of the Hungarian revolution and freedom fight in 1956 it became the garrison of the Soviet Red Army troops (4-5,000 Soviet soldiers) that were temporarily stationed in Hungary. The ownership of the Soviet barracks in Dunaföldvár was transferred to the Hungarian State in 1989, and after that one or two business enterprises made use of the site as lessees. In 1996 the Municipality of Dunaföldvár notified ÁPV Rt. (Hungarian State Privatisation and Holding Company) that it intended to obtain the site's ownership right for residential and industrial park utilisation purposes. As a consequence, in May 1997 the ownership of the former barracks and drill-ground was transferred to the municipality free of charge, and in 1998 the site officially became an industrial park. The municipality entrusted the Dunaföldvár Industrial and Regional Development public company (founded and 100% owned by the municipality) with the running of the industrial park.

During the establishment of the industrial park the operating company had to overcome several difficulties, especially in the fields of infrastructure and environmental pollution that occurred as a result of the previous Soviet military utilisation of the site. The former barracks needed to undergo a major renewal because the buildings on the site were in an alarming state of neglect, and also due to the woeful condition of the various materials, fittings, wiring installed in the facilities during the Soviet Army's stay. The replacement of fittings and wiring was also justified by the fact that during the construction and rebuilding phase completed in the period of Soviet ownership the materials employed and the way they were installed did not conform to Hungarian standards. This basically prevented the sites from being utilised for civilian purposes as they were, without making any actual alterations. The public utility networks (water and sewage systems) installed on the site were also in poor condition at the time of the take-over. Moreover, the sewage-water from the site flowed into the Danube without any treatment whatsoever.

Before the old barracks could be utilised as an industrial park, the rectification of environmental damage caused at the site had to be performed, as prescribed by law. This entailed the cleaning-up of diesel oil spills around the former petrol station and the underground oil reservoir near the central boiler house (the solution was the replacement of the contaminated soil).

The utilisation of the real asset could occur only after the above-mentioned rectification and infrastructure investments were completed. The site served two aims: on one part of the site – covering 30% of the real estate – a residential park was established (after the reconstruction of the former officers' flats and the construction of a new residential park comprised of 240 flats), while on the rest of the site an industrial park was built.

The Dunaföldvár Industrial Park is located along main traffic roads, 1 km from the cargo vessel harbour of the river Danube. The industrial park's total surface area of 32 hectares is divided into three parts: the area of the former Soviet barracks (13 ha), the former military drill-grounds (13 ha), and the areas of greenfield investments (6 ha) provided by the municipality. Currently there are 18 enterprises in the industrial park – primarily dealing with shipment-transportation and trading – which employ over 200 people.

Services provided by the industrial park for incoming residents are:

- technical counselling
- legal counselling
- financial counselling
- accounting counselling
- work force recruitment
- transportation and shipment services
- logistics and storage services
- area supervision services

4.2 Airports that remained unused in the provinces

There are more than 30 registered airports in Hungary. Of these about twenty former Soviet military air bases have been waiting to be privatised or utilised in some way or other for ten years now. Most of the currently operating facilities serve sports purposes, but some of them are fairly large airports. These include Debrecen airport and Balaton Airport in Sármellék, both of which have concrete and grass-covered landing strips.

There is practically no domestic air traffic in Hungary, although it became theoretically possible after the former Soviet military air bases were abandoned. The presence of landing strips, however, is not sufficient in itself to initiate air traffic. Airports need formidable, capital-intensive investments for which the owners (mostly municipalities or the State) do not have the necessary resources. A fierce struggle has been going on for years for the possession of the still-available old Soviet airports currently owned by the State. Experts claim that domestic air traffic and demand for domestic air transport will rapidly grow in the coming years. However, the former Soviet airports in Hungary represent potential losses for the provinces, not just because of the lack of their utilisation. In certain cases it is not enough that the facility once operated as an air base – which was secret and, as regards its infrastructure, was considerable; it has been deteriorating for ten years now. Owing to its presence, the development of the settlement itself could be delayed for decades.

The history of the **airport in Sármellék** goes back to the most tense years of the Cold War. The airport in Zala County was originally built by the Hungarian Army. The area covering 408 hectares was later handed over to the Soviet troops that used it up to 1990 (during the last year the 5th Guards Fighter Regiment with MiG 29 (Fulcrum-C) and some MiG 29 UB's was stationed in Sármellék).

The expected Western attack did not occur, and after the political transition both Western and Eastern planes utilised Balaton Airport. Instead of soldiers, however, the charter flights arriving from Germany, Switzerland and Russia transported tourists. The volume of business and tourist civilian air traffic by light aircraft has also been rising steadily. Developed from the former Soviet military base, Balaton Airport has a favourable location as it lies 12 kilometres away from the prominent tourist centre, the "Hungarian sea", i.e., Lake Balaton, 11 kilometres away from Hévíz, and 22 kilometres away from Zalakaros, the most prominent health spa resorts. In 2000 about ten thousand passengers landed at the airport in Sármellék. In 2001 the owners anticipated a doubling of the previous year's traffic volume due to the opening of the customs control at the airport, so it was not necessary for the arriving passengers to obtain a landing permit days in advance.

The government made a decision about the future of the airport in Sármellék in November 2000. It decided that, in addition to the Debrecen air facilty, this airport should be developed into a regional international airport over the next few years. This would be necessary because there are just two countries in Europe, namely Albania and Hungary, which have no international airports in their provinces, only the ones in their capital. According to the decree the settlements operating the airport – the municipalities of Zalavár and Sármellék – shall obtain the facility free of charge, but shall assume all other costs. Although the construction of the infrastructure essential for the safe operation and traffic increase at the airport in Sármellék represents only a relatively small part of the total costs, even according to the most conservative estimates, billions of forints will be needed to complete this part of the project. The major portion of the costs, however, lies elsewhere. In contrast with many other former Soviet military facilities in Hungary, nobody has really assessed the extent of the environmental pollution caused, although probably the airports present the gravest environmental issue. There are no exact figures available on the estimated costs of environmental rectification as no damage assessment has been done so far, but experts unanimously claim that the costs of environmental rectification will be substantially more than the sum required for the above-mentioned infrastructure investments.

The establishment of **Debrecen airport** dates back to the 1930's. Although originally it was designed to assist the civilian air traffic and transport service, during the Second World War it served the purposes of Hungarian national defence. The Soviet Air Force appeared in Debrecen in 1946, and was stationed there until 1990 (during the last year the units stationed in Debrecen were the following: the 88th Fighter-Bomber Regiment with 38 MiG 27D and MiG 23UB, and an unknown EW squadron with Mi-8SMV/PP/T).

After the withdrawal of the Soviet troops the airport was transferred back into the hands of the Hungarian State. The municipality of the city purchased Debrecen airport in 1993, and up until now it has spent nearly HUF 100 million on developments. Apart from the runway itself, the departing Soviet troops did not leave much behind. Although the replacement of the safety equipment was costly, the attitude of the owners is upbeat. The city plans to open the airport for international civil aviation with high hopes, because they think that many foreign investors seek settlements that can also be approached by air. Eastern Hungary has identified this demand and considers that the airport may boost health tourism and thus it could become a key player in the region's economic life.

Since May 2001 Debrecen airport – apart from the airports in Budapest and Sármellék – has also functioned as a customs control, and during the summer of 2001 scheduled flights arrived from Vienna. By opening the Debrecen customs control, it is no longer necessary for planes to ask for landing permission 72 hours in advance. Despite these measures, however, the airport still does not conform to the standards set by the European Union. The terminal building has to be reconstructed in line with the Schengen regulations, and the technical background – ensuring the take-off and landing of the planes in all weather conditions – also needs to be set up.

Debrecen Airport is the second biggest airport in Hungary. The 2,500m. long concrete runway is capable of receiving even the biggest passenger airplanes.

Taking into account the local conditions, it is no wonder that the municipality of Debrecen has been making every effort to sustain and further develop the airport. Moreover, Ferihegy airport in Budapest has no diversion air-field, so the flights that Budapest cannot receive are bound to fly to Vienna. Debrecen would like to fill this gap. After the realisation of these airport developments, Malév (Hungarian Airlines) will presumably use Debrecen Airport as a diversion air-field instead of Schwechat.

Unfortunately, the resources of the municipality were too scarce for them to operate the airport on its own, so Debrecen then had to look for co-investors. The former Soviet airbase near Hajdú County's capital was purchased by a Turkish real estate investor company named Begendik from the municipality of Debrecen. The buyer undertook to develop Debrecen Airport into an international airport within three years and to perform the needed environmental rectification as well. Depending on the level of the environmental damage caused, this will consume a sum between HUF 200 million and 1.2 billion. The municipality undertook to provide assistance to the Turkish party in preparing the tenders in connection with the environmental rectification. The Turkish party is expected to become the owner of the said airport after paying the last instalment in 2002. The buyer, the Begendik Holding Company, is also interested in the airport in Sármellék. The operators of this airport consider the said company as the only serious investor.

The two examples mentioned above provide optimism for the future of Hungarian civil aviation. According to expert estimates, however, a two hundred and fifty thousand-strong passenger volume is required for the profitable operation of an airport. There seems little hope that in the next few years any Hungarian airport outside the one in Budapest could really meet this target. Until then the present situation is likely to remain: investor apathy, the lack of State and EU funding, and a general resistance of the population to even the smallest developments.

4.3 Officers' flats in the service of the civilian population

In Hungary, one of the commonest ways of utilising the former Soviet military facilities is the reconstruction, selling or leasing of real assets for residential use. The solution seemed quite obvious. Over the past decade the number of new flats being built has been on the decline, so there was a growing demand for new flats, especially among newly-weds. Consequently, the Government and the municipalities placed great emphasis on the stimulation and support of private investments as well as the realisation of investments aimed at increasing the availability of rented flats in settlements.

Here the municipalities considered the use of former officers' flats and unused areas of former Soviet barracks for residential purposes an excellent means of utilisation. The civilian utilisation of the 15,000 flats nation-wide was realised, although it was hampered by the already well-known issues. A comprehensive renewal of the buildings was necessary, in most cases only the structure remaining unchanged. New solutions for the public utility networks, pavements, heating systems, roofing, and often the total heat insulation (along with the replacement of doors and windows) were also found necessary. All these significantly augmented the costs of their reconstruction. However, as luck would have it, the environmental pollution issues outlined above did not affect these residential parts of the sites. Due to this, the overall cost of building new flats is comprised of the costs of refurbishment, the construction of public road networks, pavements and playgrounds as well as landscaping of the residential surroundings.

The renewal and utilisation of former Soviet flats in Hungary was realised by different models of financing. A good illustration is the conversion of the former Soviet barracks in Debrecen into a residential park. Here the municipality sold a part of the area to investors while it refurbished the other part from its own resources with the aim of creating flats for welfare purposes. Within the framework of an investment initiated in 1998, the municipality spent some HUF 1.3 billion on turning the buildings – which had been abandoned and in bad condition after the withdrawal of the Soviets – into habitable residences. Seven residential buildings underwent a full-scale renovation, which was followed by the infrastructural development of the residential park. During the reconstruction of the 366 flats located in prefabricated buildings only the structure remained unaltered. New building engineering, electrical, water, gas and heating systems were installed. The wall and floor coverings had to be replaced and the roof structures and building insulation had to be likewise repaired. During the development the full range of public utilities was installed on this site. Another vital task was the construction of public roads and pavements. The refurbishment of five other buildings also located on the site of the former barracks was realised by the sale of the buildings to an investor and, as a result, a further 70 flats were made available.

With the aid of a bank investment, a new residential area was built in Veszprém on the site of a former Soviet barracks. It has a favourable location, so the town was made complete with this popular residential area. In 1997 OTP Ingatlan Rt. (OTP Real Estate Plc) concluded a long-term co-operation agreement with the municipality of Veszprém. As a result, the so-called Kiev-type buildings were reconstructed one year later. In 2000 OTP Ingatlan Rt. invested one billion HUF in the construction of flats in Veszprém: it purchased new plots, equipped them with public utilities and, by the end of November 2000, it handed over a block of 44 flats costing HUF 600 million.

There is a strong demand for the former officers' flats, partly as a result of the decrease in flat construction in the past few years, and partly because the so-called Kiev-type prefabs have come a long way and can by no means be compared to their former state. As a result of reconstruction, the walls were covered with a heat insulating material, the staircases and entrances were redesigned and their appearance is much improved. Looking at these renewed flats from within and without, it is hard to believe that previously they once had a military function and that the prefabricated panel-structure buildings had been mass-produced during the socialist era.

4.4 The utilisation of the abandoned real assets for higher education purposes

In trying to realise the political ambitions of higher education (i.e. raising the ratio of the students participating in higher education to 50-60% of the characteristic age-group), the higher education institutes were faced with the problem of establishing the necessary infrastructure background for ensuring the conditions of learning for students whose numbers are already high today, and are set to grow in the future. After being well renovated, the abandoned Soviet and Hungarian military barracks provided an excellent way of enlarging higher education institutes, being utilised for educational, campus and dormitory purposes.

The **Kossuth Barracks in Szeged**, which was decommissioned during the course of the Hungarian defence reform has, over the past decade, provided one pillar of development at the University of Szeged. The faculty of the Teacher Training College acquired buildings from the Hungarian state in three phases. First, in 1992, the building formerly used as living quarters for military personnel, then two additional buildings in 1995, and finally another taken over by the university in 1998 where further departments and lecture halls were established.

Another important asset shaping the development of the University of Szeged is the site of the former Soviet military barracks on the outskirts of the city. In 1994 the University was granted ownership of the site, which had been in the hands of the Hungarian State from the early 1990's. According to the original plans, a university campus was to be established on the site. Nevertheless, taking into account its history and the likely future opportunities for higher education in Szeged, in 1995 the leadership of the city and the University reached a crucial decision. The parties agreed that the University, mostly located in the inner city area, would not move to the site of the former barracks. Thus, they decided to exchange the real estate acquired by the University for other inner city real assets that were in the ownership of the municipality. Between 1995 and 1997 no investment was made on the site as the University and the city could not reach an agreement about the final form of utilisation, which was, among other things, hindered by the issue of the site's environmental rectification. Meanwhile, several concepts emerged, such as its utilisation as an economic zone, industrial park, or scientific park. Finally, in 1997 the utilisation issue was solved: the former officers' flats and the former community arts centre remained in the hands of the university, while the other areas were offered for sale or leasing by the municipality for economic, mainly innovation utilisation purposes (the primary objective of utilisation is the establishment of a scientific park).

As regards the utilisation of the community arts centre, which is located on the site of the former barracks, advanced negotiations are under way, and the University has a vested interest in creating a biotechnology incubator. The first phase of the investment, during which the former officers' flats were turned into a hall of residence, was completed between 1997 and 1999. By renovating one of the three buildings at the site with a budget of HUF 270 million, a hall of residence accommodating up to 270 students was created, and is now up and running. As far as the University is concerned, this investment will not mean the end of attempts to create additional accommodation for students. By refurbishing the other two buildings, the University plans to make available a further 60 + 60 flats, that is, altogether 480 beds in the near future. But up until now the problem of how to secure the funds needed for the renovation (the costs of which are rather high, namely HUF 1-1.5 million per bed) has not been overcome.

The developments outlined above have enabled the University to double its student intake over the last decade. At present the University provides higher education training for some 27,000 students.

The Threats and Opportunities of the Changing Supply Chain in the Aeronautics Sector

Brian MORRIS
AECMA
Brian Morris is the Equipment Attaché at the European Association of Aerospace Industries (AECMA). His responsibilities include representing the Equipment Sector with EU and European Parliament and other institutions. He is also heavily involved in Supply Chain issues, particularly relating to SME's. Prior to taking up his role at AECMA he was Commercial Director of Marconi Aerospace Systems.

Abstract. There are around 8 million SME's in the European Union, as currently constituted and the addition over a period of time of the New Accession States will bring in many more. SME's are a politically hot item both in the European Commission and the European and National Parliaments. They are innovative and are crucial to the well-being of European Industry.

AECMA is the European Association of Aerospace Industries. It represents the aerospace industry in Europe in all matters of common interest with the objective of promoting and supporting the competitive development of the sector.

The 16 members of AECMA are the national aerospace associations of all EU Member States and the Czech Republic, as well as the largest European aerospace companies; discussions have also taken place with Poland and Turkey as prospective members.

1. Key figures of the European Aerospace Industry

Companies: about 760 with a further 80,000 suppliers, many of which are SME's.
Employment: more than 429,000 people direct and about 1.2 million including indirect employees.
Turnover: more than 72 billion Euro. R&D expenditure runs at around 15% of turnover.

I will today deal primarily with the Aeronautics Industry. The area within Aeronautics I represent is the Equipment Sector, which represents around 30% of the turnover of the Industry (72 billion Euro). The SME's in the Aeronautics sector are faced with a rapidly consolidating industry.

The spread of activities of the Prime Contractors/Systems Integrators is wide, and with the number of joint ventures and partnerships increasingly complicated. Fig.1 tends to illustrate this.

How can SME's deal with these changes and what are the threats? What are the opportunities?

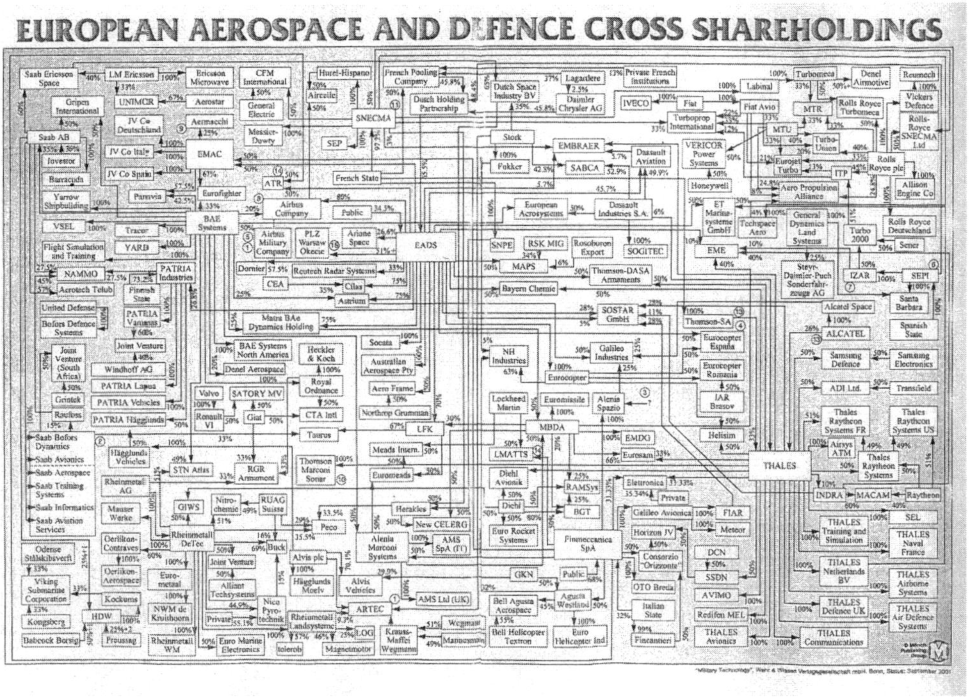

The Aerospace supply pyramid now looks something like this.

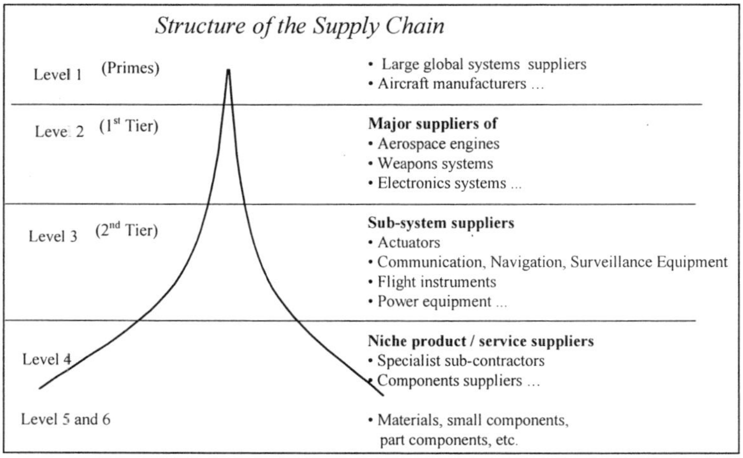

The 1st Tier Suppliers have also reorganised to take account of the devolved responsibilities passed down to them by the Prime contractors and many of them are now system integrators in their own right. These 1st Tier system integrators are the best targets for SME's, as the Prime contractors demand products and processes that are integrated at the platform level. This downward push of risk and responsibility has an effect lower down the Supply Chain where small companies find themselves squeezed on time, on cost and frequently (if you ask them) information on long-term planning.

This particular issue is not a new one. The proper functioning of a flexible responsive Supply Chain must be the aim of all customers and suppliers in that Supply Chain.

Aeronautics is a global market and many European suppliers compete globally and have overseas Primes as customers. It is interesting to note that depending on the mix of engines and equipment, over 40% of an Airbus can be provided by the US (and other) suppliers.

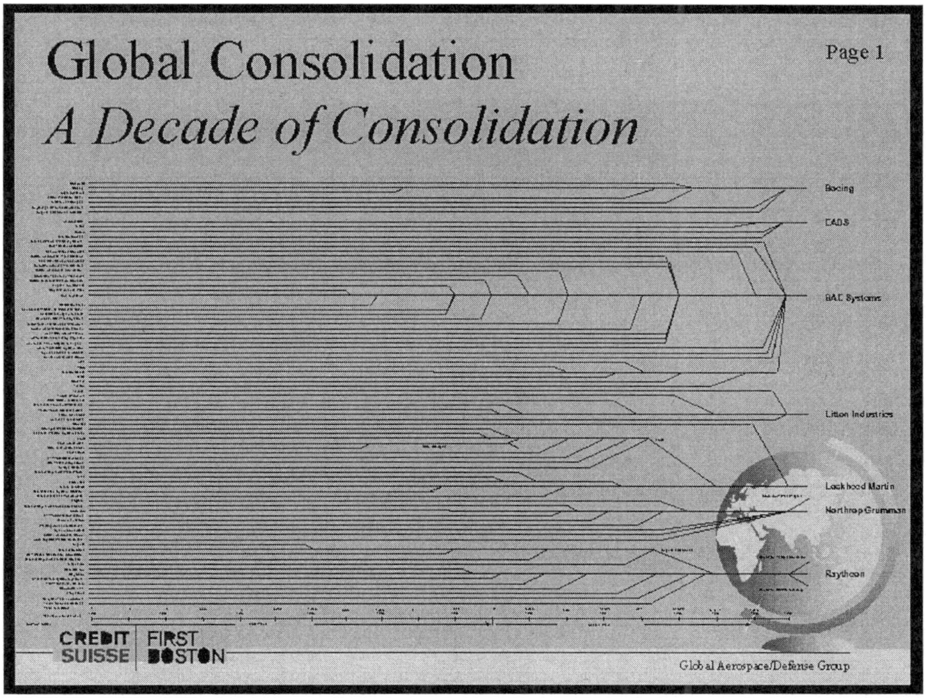

Where does this leave the small supplier? Unless that small supplier is proactive it will leave it outside of the aeronautics Supply Chain in the future.

Let us now look at some of the threats and opportunities facing small companies.

2. The threats

The dividing line between business segments is becoming blurred. There are proactive companies seeking new opportunities outside of their traditional business markets. The evolving use of e-business/commerce means that traditional Supply Chains are under the additional threat from new methods and approaches. The drive for fewer suppliers which is the traditional yardstick for the procurement department is being looked at in the light of evolving technologies on the web, where large companies can access other portals and industries in order to seek out non-traditional suppliers, who may not only have cheaper products and/or processes but innovative ideas and experiences. Nevertheless there are still barriers to new entrants based on qualification and the ability to demonstrate robustness for the long haul. In addition Primes are increasing their ability to set up new suppliers in place of those that have failed their regular assessments of performance.

However the messages here are: decreased number of suppliers, increased turnover per supplier, modular design of products leading to less parts, reduced time to market and increased technological demands.

An important statistic for EADS Airbus Germany in 1999 was that 3% of its suppliers accounted for 80% of the purchase order value. Experience in the automotive industry has demonstrated the same effect; number of suppliers down, spend per supplier up.

We are competing in a global market. The new emerging countries such as Brazil, China and India offer a threat. They are competitive, have access to first class technology and partners from around the world. In Europe as the Industry restructures and develops, the question of support for R&TD and so-called launch aid is an important area for consideration.

The European Aerospace Industry has currently and in the past seen major R&TD support from the European Union. In the Fifth Framework Programme some 700 million Euro has been allocated to the Aeronautics sector. This has allowed major long term Technology Platforms and Critical Technology Projects to go ahead on a 50/50 basis with the Industry. The history of success in the Aeronautics sector had been around 3% in previous Framework Programmes. The reasons for this are many, but the key ones are that the Aeronautics sector is highly structured and long term. making it difficult for small companies to undertake the heavy financial, and long-term programme commitment required by the major Technology Programmes. Also remember that EC funding is not cheap money and in R&TD the SME has to provide 50% of the funding and share the risks, with no short-term rewards. Active efforts by the EC and Industry have in the 5th FP increased SME participation to some 6%, and this also takes account of medium-size companies which do not fall within the EU definition which is aimed at CRAFT projects.

In the new Framework Programme 6, the increased target for SME's is 15%. This means that SME's would have to raise 15% of 1,075 million Euro. This is where the political aspirations come face to face with the realities of the real supply chain that exists.

The R&TD monies of 20 years ago and the repayable launch aid have turned into the products of today. This of course applies in the field of defence-related R&TD which remains firmly based on a national basis. I will return to this a little later.

While I have said that dealing with the EU means risk sharing and investment, this has been an increasing aspect for companies in the Supply Chain. Risk and revenue sharing is now a common approach; suppliers face investment in non-recurring costs, provision of no-cost products and no reimbursement if products are not proceeded with or replaced.

On top of this the Primes are increasingly faced with the price a customer is prepared to pay, and this target is shared with its suppliers. For suppliers in the Supply Chain there is continuous assessment of performance, and there may well be alternative suppliers waiting in the wings.

Let us look at the US approach and the European approach to smaller companies in all sectors. The USA is ahead of Europe in allowing SME's to start, grow and function. Bankruptcy does not have the stigma or the legal restraints that it does in Europe.

There is a great deal more money available in the US than in Europe for start-ups and maintenance of growth of SME's; around $50bn in the US against $14bn in Europe. Of that $14bn the UK provides around 50%. The US GDP is 4 times larger than Germany's; its venture capital is 30 times larger.

The moves being made to introduce a European patent are to be welcomed. This should assist small companies. Currently it is twice as expensive and takes five times as long to set up a patent than in the US.

E-commerce is here and the EU must push through new laws which will enable European companies to be able to operate effectively.

These issues on a wider front covering all trading sectors are being addressed by the Trans Atlantic Business Dialogue, this year being co-chaired by Phil Condit of Boeing and Sir Charles Masefield of BAE Systems. For the first time the SME view has been included as a priority and AECMA and UEAPME are actively involved.

I have said before that the USA is not just a competitor. It is for many companies a major customer and vice versa, although the two-way street in the defence area may be a little more one-way than many would like. European companies must be able to compete in what is a global market and an effective Supply Chain is a key factor in maintaining that competitiveness. However, the difficulties of an SME operating in a multinational project are not to be underestimated. This is a particular difficulty in EC R&TD projects where multinational groupings are a prerequisite.

On the defence side, spending (with one notable exception) is being reduced and, now within Europe, companies face changing Government Procurement policies; the creation of OCCAR is but a first step; but how far will the European Industries be allowed to rationalise by their respective governments, some of whom may wish to maintain what are regarded as strategic assets?

Recent operational experience has demonstrated that the expenditure on European defence does not provide a proportionate operational capability compared to the USA.

The subject of civilian versus military R&TD in the EC Framework Programmes is an issue that is worthy of discussion, but not here.

Where does this leave the SME who is more concerned about micro rather than macro events? Contracts and cash flow are the major concerns of such companies.

3. Clusters of SME's in European regions

Associated with Supply Chain and SME issues is the regional structure of the Aerospace Industry. The European Aerospace Industry has activity in all the 15 European Union member states as well as other European countries. Concentration is evident in certain countries and within these countries/companies are concentrated in certain areas. Examples of the clustering effects are well known within the aerospace community. One of the best known in Europe is Toulouse in France, due to the Airbus programme.

In the UK, BAE Systems and Rolls-Royce facilities have acted as a source for the generation of associated businesses.

Germany, Italy and Spain also have their aerospace clusters.

One developing area is the clustering together of SME's in an informal or formal alliance. These groupings usually offer a range of services and/or products. This enables them to deal directly with the large players. The co-ordinated use of staff and facilities allows a much more effective deployment of resources. DevAero* in The Netherlands and AATG* in the UK are examples of such clusters; a recent presentation by AATG made the following points. By working in partnership as a trading group or consortium SME's can bid for programmes of work they would not otherwise have been able to consider. Through such partnerships manufacturing SME's will have greater access to design capability within the SME population and thus an opportunity for added value and a move up the Supply Chain. The relationship with academia should also be developed so that the aims of Industry and universities can be better directed and co-ordinated.

4. The opportunities

Now let us look at some of the opportunities. As I have said previously, perceived wisdom is that there will be less suppliers, bigger suppliers; that in general is true and this process is ongoing in European Industry.

In the USA, where the re-organisation of the Aerospace Industry has been in advance of that of Europe, it has been found that proactive small and medium-sized companies have been able to increase their share of business as the big companies become less agile and seek solutions and products which they themselves are not structured for or minded to produce. Opportunities will exist to provide a product to a range of companies as the supplier of choice or to provide a wider range of products so that not all the eggs are in one basket. As I have said previously it is in this area that co-operation between smaller companies can develop as a virtual company which has a formal or informal association which lends it size and capabilities that larger companies can/are prepared to deal with. It is important for a company to decide its place in the Supply Chain. It is here where the 1st and 2nd Tier Suppliers are important as it is these companies that have the most inter-relationship with the smaller members of the Supply Chain. The ability to share risks, financial viability, flexibility and quality are all givens in this market. What is the differentiation of an SME? With the restructuring of the Supply Chain there will be strategic suppliers who will be a partner or preferred supplier; those that are being developed, new suppliers and finally those with limited use and those that are cancelled. It is where a company as a best practice supplier finds itself on this list that may well direct its objectives and aspirations. The three most important words that apply to the operation of the Supply Chain are visibility, openness and trust.

Finally the advertisement for some practical free of charge assistance. The AeroSME website at www.aerosme.com gives access to an Aeronautics database of over 500 SME's, as well as Regional Associations and Research Institutes. This database is set up to facilitate networking and partnering for proposals into the EU's Sixth Framework Programme and as a joint EC/Industry activity has involvement from 20 countries, and is being extended to cover the New Accession States. The first SME's from Latvia, Estonia and Romania are already registered. I should point out that registration on the database is limited to SME's that are from countries that are contributors to the Framework Programmes. However the use of the database for information and networking by non-qualifying SME's is available free of charge by simply clicking onto the AeroSME website. In addition, our role is to facilitate contact with other EU SME initiatives and SME-related activities in other industrial sectors. Any SME is welcome to join the database by filling in a questionnaire which is also available at the AeroSME website.

It is an important point to note that many of the SME's on the database are not solely aeronautical. Many have business in defence, shipbuilding, railways, automotive IT, etc. The aspirations for defence-focused companies to extend into another sector can be supported by AeroSME. The rules for the 6th Framework Programme are clear in that defence R&TD is not directly supported. However the application of skills from defence SME's that meet the civilian research criteria is an area of perhaps untapped potential.

What should SME's be doing?

Finally, let me address some of the micro things that SME's should be getting on with.

The key to success and the maintenance of a business is knowledge. SME's cannot control the macro events but the proactive and successful ones have objectives, and above all work at achieving marketing knowledge and make networking an essential part of their activities. A proactive SME will understand and define its place in the competitive framework. It must then identify its key success factors and revise its organisation and skills accordingly; it must benchmark and understand the standards for success; it must then invest in those areas that will bring success. Only by seeking a competitive advantage will any SME be able to maintain its position in a rapidly changing world and a Supply Chain that is changing so fast that it is leading to new relationships for all companies from the top to the bottom of the Supply Chain.

Although I have spoken about small companies we should not lose sight of the small countries where the development of key skills applying to the Aeronautics sector is again an area which AECMA via AeroSME provides assistance and advice.

The sharing of knowledge is one of the most effective ways to help the SME community to help itself.

Links

www.devaero.com
www.aatgl.com
www.aerosme.com
www.aecma.org

Defense Related SME's: Analysis and Description of Current Conditions
F.D. Carvalho (Ed.)
IOS Press, 2004

The Relationship between SME's and Prime Contractors

Guy SERVOLLE
Corporate vice-president Dassault aviation: European Union and NATO, France

Abstract. This paper presents an example of structural elements' production with partners. I remind you of the well-known context underlining what seems to me the most important things for our industry in an extremely unfavourable context, at least as long as the analysis of our new threats has not yet concluded in adapted armaments. It is up to the prime to implement the support of his commitments and to create a real partnership with the chosen sub-contractor.

Offsets "work", but we have to prevent any unjustified commitment and to make all relevant efforts when the contract is signed.

To optimize the cost of the structural parts for our aircraft we must work together as early as possible in the definition cycle to benefit from the knowledge of the industrial means which will be used for the industrialization and production phase.

First our choices between "make or buy" are based on what we want to keep as our core business in order to be able to deal better with our production network and to develop the necessary technologies with the means of a large enough company.

Introduction

Let me first thank General Castelo Branco who gave me the opportunity to present to you today the experience I acquired when working for Dassault-Aviation.

I was indeed in charge of the so-called "External Production Department" from 1984 to 1990 and was responsible for our relationships with sub-contractors from large to small and medium-size companies. In preparing this presentation I updated my knowledge with my successor!

To be clear, I confirm that I would like to address the partnership with aerostructure manufacturers for my company, the design being made by Dassault, the definition being more and more detailed by the sub-contractor himself.

Concerning context, this session refers to a previous period, in principle less constraining than the current one, but there are for us no significant differences between our past and present relationships. Therefore I would just remind you of the well-known context underlining what seems to me the most important things for our industry in an extremely unfavourable context, at least as long as the analysis of our new threats has not yet concluded in adapted armaments.

Everything is unfavourable for us in the means by which our European States invest in our activities, when our shareholders are looking for profits and our institutions are developing the ideologies of the "acquis communautaire".

1. Offsets: "they work!"

I do not want to speak about SME's, particularly about those abroad, without presenting my views on the offsets question, for which lots of wrong things have been said by those who believe that their interest is to cancel their practices.

It is obvious that nobody from the selling countries wishes their continued existence, but everyone continues to propose them when it helps to support a proposal (see for example what is usually said by the US and what they promised to South Korea in the deal for fighter aircraft).

All that we can try to do is to organize them better than we did before. No commitment should be taken into consideration if the salesman is unable to demonstrate in detail which partner he will work with and what the work packages are that he will order from him. It implies, and it is what I did in Greece and in Egypt for example, a detailed evaluation and the implementation of an industrial relationship before any commitment. Furthermore, to avoid that work does not progress properly when the sub-contractor is unable to respect schedule quality or price, the termination rules have to be agreed by all parties.

It is then up to the prime to implement the support of his commitments and to create a real partnership with the chosen sub-contractor. For example, in Greece where production of machined parts was not progressing satisfactorily, we seconded a foreman to advise on necessary investments and to organize a workshop in order to achieve the expected production ratio.

My conclusion on offsets is that "they work", but that we have to prevent any unjustified commitment and to make all relevant efforts when the contract is signed.

2. Needs to optimize costs

That being said, what are our needs in optimizing the cost of the structural parts for our aircraft? We must work together as early as possible in the definition cycle to benefit from the knowledge of the industrial means which will be used for the industrialization and production phase. It implies sharing the same digitized databases and the same CAD-CAM tools in what we call "entreprise élargie" (enlarged company).

No more mock-up, sample parts and interchangeability jigs; digitalization replaces everything, provided we avoid any translation in different systems which is generally a source of loss of information and even errors.

Furthermore, we add to the concept of "entreprise élargie" the concept of working "en plateau" (28 companies, 8 nationalities for the new business jet Falcon 7 X), not only to enjoy the concurrent engineering but also to let the partner detail the definition of his WP from the commonly-developed principles and no longer from a specification written unilaterally by the prime alone.

Cost optimization as required and supported, such as is described above, can only be achieved by a total agreement with the partners and provided that their competencies and stability enable them to do so. Furthermore, we make a continuous effort for market changing due to economic conditions and international shocks. We must permanently adapt ourselves to satisfy our customers' needs. It draws my attention to the fact that I am not making any difference between the civilian and the defence business. Really, even if the customers are totally different, the relevant methods of partnership are the same!

In both cases, the most recent technological developments have to be implemented and it is a must for our partners to keep themselves at the right level with the fastest reactions.

Industrial work-sharing can also be modified depending on alliances/mergers and the wishes of our customers.

It leads to mandatory clear procurement policies to face any adaptation to a fast-changing world.

3. How we try to manage our supply chain

First our choices between "make or buy" are based on what we want to keep as our core business in order to be able to deal better with our production network and to develop the necessary technologies through the means of a large enough company.

Our priorities once defined, we face a difficulty in the choice of our partners:

- On one hand to reduce their number in order to know them better and to integrate them in our "entreprise élargie"
- On the other hand, to open the competition against our usual partners to stimulate their competitiveness.

I do not know any perfect solution to this dilemma, but pragmatism and a case-by-case basis supported by some criteria help us to proceed. Within these criteria let me mention some of them such as:

- The division in 3 classes depending on the level of involvement in the definition and on the size of their work packages (WP);
- The turnover realized with us, which we are pleased to see at around 30%; enough to be an important customer, but not too much to minimize the impact of possible reduction in our common activities.

Moreover:

- As close a relationship as possible with the factory using the WP on its production line to avoid any bureaucratic interface and to solve any potential problem as soon as possible.

Concerning the content of the sub-contractor WP, it has to be as flexible as possible to adapt itself to the sub-contractor's capacities and as precise as possible to avoid any misunderstanding between partners.

Therefore we need to write together a detailed specification that clearly describes the limits of proprietary rights and the statement of work.

For example, it could be considered as efficient to let the partner provide the raw material, nuts and bolts for his WP or, on the contrary, to benefit from the scale effect of having the prime contractor doing that work.

I mention this opportunity only to insist on the absence of rigid rules and to emphasize the possibility of adapting choices to each other's capacities.

This is really the main criterion for the success of our partnerships; the continuous adaptability to the ceaseless evolution of our industrial environment.

I thank you very much for your attention.

Dual Technologies – Basis of Innovation Development of the DIC enterprises of Russia

Vladimir SHABAROV
Russia

Abstract. Innovation policy is an important component of Russian Government social and economic policies.

It is necessary to note that in the Russian Federation, as well as in a number of other countries, there is an unequivocal definition of the term "dual use technologies". Usually this term – dual use technologies (DUT) is employed to define technologies both in civil and military industries.

As a rule, military technologies are science-intensive ones. However, civilian products made in Russia with the use of these technologies often prove to be uncompetitive. The reason in many cases is that the goods and products manufactured according to these technologies are too expensive. Why? There are several reasons for that.

Despite the obvious advantages resulting from introducing DUT, first of all in the sense of opportunities of receiving fast feedback from financial investments, there also exist certain difficulties in their introduction.

The implementation of innovation policy in the field of DUT taking due account of the potential and interests of all innovation activities' subjects and international, economic, social and political priorities, will promote the reorganization and rise of the national economy of Russia, its successful cooperation with Western countries and its integration into the world market economy.

Introduction

Innovation policy is an important component of Russian Government social and economic policies.

It is designed to determine purposes, strategy and mechanisms of supporting priority innovation programmes and projects. The State's objectives in this area today are:

- The increase of efficiency of using scientific data information for practical introduction of results of fundamental and applied research into production.

Fast regression and structural crisis, which have been reigning in the Russian Defence Industrial Complex (DIC) since the beginning of the last decade have strongly reduced its potential. Simultaneously, Government investments and other forms of State support to scientific research and development have strongly diminished.

The direct result of this has been a sharp decrease of production volumes of science-intensive produce, which has been for decades the basis of economic development of the country.

Today's dilemma is the following: either the loss of DIC industrial potential will be compensated at a modern scientific and technical level, which will require a sharp increase in investment and innovation activities, or the country will be thrown several decades back not only in production output, but also in the technological field. The answer in our opinion lies in the field of applying such methods and such technologies that would assure the introduction and efficient use in production of scientific and technical potential. The so-called dual technologies are the most important component of such technologies.

1. Problems of terminology

Its introduction and use can certainly positively influence both private small and medium public enterprises (SME's) and organizations.

Usually this term – dual use technologies (DUT) is employed to define technologies both in civilian and military industries.

It is necessary to note that in the Russian Federation, as well as in a number of other countries, there is an unequivocal definition of the term "dual use technologies".

However, there are also some subtleties that are reflected in the appropriate definitions in the narrow sense of the term.

- In accordance with "Provisions on the control over carrying out obligations on guarantees of imported and exported goods (services) of dual use according to the declared purposes" (Decree of the Council of Ministers of the Russian Federation No 1030 of 11.09, 1993) the following definition of dual use is employed - "the goods (services) of dual use are equipment, materials, raw materials, technologies and scientific and technical information used for civil purposes, but can be applied for producing arms and military equipment, including weapons of mass destruction and missile means of their delivery". We shall conditionally name this definition "definition No 1".

  Traditionally (and most frequently) this term is used in Russia in connection with the system of exports control.
- In definitions used for the objectives of conversion, priorities are placed differently, according to the return order of the relationship "primary – secondary" ("producer – product"), as "military technologies, which can be used in manufacturing civil production" (definition No 2).

  By dual use introduction we shall mean in this case organized manufacturing production of civilian production which is economically profitable.

  The introduction of DUT in the sense of definition 2 is profitable to the enterprises of the defence industrial complex in many ways. Firstly, the strategy of employing DUT allows the enterprise quickly and with minimal losses to switch to civilian conversion production (in case of absence of the State defence order), and, secondly, in doing so to preserve its mobilization readiness of fast restoration of its military potential.
- In processes of post resource (post-guarantee) DUT are sometimes understood as technologies of using available (or partly modified) DIC products, commissioned from armaments after the expiration of the specified service life or according to programmes of disarmament or re-equipment, but suitable for use in civilian industries. Examples are use of air engines that have outlived the specified service life, but are virtually operational according to their technical condition, which can be used for problems of icing or for use in the gas industry, use of the chassis of tanks as a basis for cross-country vehicles, etc. This direction is relatively promising in connection with reduction of the armed forces and implementation of disarmament programmes (definition No 3).

  It is obvious that the first definition does not fall under the theme of the report.

  Under the innovative development of the DIC enterprises we mean the process aimed at implementing results of completed scientific research and development or other scientific and technical achievements, in a new or advanced product or process, sold at market or used in practical activities. It is necessary also to take into account the process of introducing DUT and additional scientific research and development (R&D) connected with it.

The innovative development process in the area of DUT can stimulate changes both in the structural and organizational sphere and in final products, having a positive effect on DUT themselves.

2. Influence of DUT on perfection of organizational forms and reforming of DIC enterprises

As a rule, military technologies are science-intensive ones. However, civilian products made in Russia with the use of these technologies often prove to be uncompetitive. The reason in many cases is the goods and products manufactured according to these technologies are too expensive. Why? There are several reasons for that. Firstly is the inclusion in the structure of price of the civilian purpose products (CPP) of non-recurring expenses for the civil reorientation of production and conversion. Secondly, a lack in DIC enterprises of flexible structures and techniques, which would allow for a quick switch to new products. Therefore the very fact of introduction of dual technologies results in a drastic change of the traditional structure of enterprises.

1. The introduction of DUT stimulates the change of an enterprise structure. Russian DIC enterprises traditionally used the so-called, linear, linear-staff and linear-functional structures of organization and management (the terminology used here comes from the general theory of management). These structures are conservative enough; it is difficult to reorient them to new tasks of market economy. The presence of these structures assumes a significant overhead charge. The introduction of DUT focuses DIC enterprises in new types of production organization. In particular, now at a number of enterprises, variants of structures adapted to changes of external and internal environment are being created. In this context we can observe the transition of enterprises from a mechanistic type of internal organization, with management based on formal rules with systems of centralized decision-making, narrowly defined responsibility and rigid hierarchy of authority, to the so-called organic structure, weakly or moderately using formal rules with decentralization of authority, with structures characterized by an insignificant level of hierarchy and wide participation of employees in decision-making.

2. A number of enterprises employ for dual use production a so-called divisional (project) structure of management. And divisions are set up accordingly both by the final producers and by the consumer and the sales markets.

3. Decentralization of DIC enterprises. In division (project) structure of management some departments of DIC enterprises, focused on a certain final product or certain type of consumer receive legal rights giving them certain financial and legal independence.

4. Stimulating and encouraging the DIC enterprises. It is obvious that all the above-mentioned innovations stimulate the development of small business and private enterprise. For the enterprise it is economically justified and profitable. According to the data of the US National Science Foundation, the number of innovations in small companies per unit of cost, as a rule, is more than in medium and large companies. Besides, the small companies almost by one-third outstrip the large ones in the speed of innovation introduction mastering. That is why within DIC enterprises the appropriate small enterprises focused on DUT are created. And in the setting up of these enterprises both DIC enterprises and the administration of the enterprises as well as the DIC in general, are interested. The reason is evident - provision of employment for highly skilled labour and also significant financial

support which these enterprises can receive from the activity of small enterprises. This process is encouraged by the prime mother companies themselves as well by the State local authorities.

5. It is obvious that the introduction of science-intensive DIC production in civilian industries and especially in the raw materials sector will allow a considerable increase in the level of production in these sectors (see examples in definition No 3).

6. The DUT development at enterprises brings about far-reaching consequence changes in the very structure of an enterprise. DIC begins to fit into the context of market relations that guarantee Russia's integration into the global market economy. In general civilian industry acquires a more science-intensive character and becomes more receptive to the latest Western technologies.

3. Some ascertaining conclusions and measures offered for the development of innovative potential in the field of DUT

1. Despite the obvious advantages resulting from introducing of DUT, first of all in the sense of opportunities of receiving fast feedback from financial investments, there also exist certain difficulties in their introduction. Companies basic difficulties in introduction of DUT innovative development are connected with limited financial resources, both budgetary and non-budgetary, including borrowed and attracted resources. Production recession in all branches of the defence industry and constant lack of investment do not often leave room for innovative activities in the field of DUT. Under conditions of deficiency of their own financial resources as well as the limited character of State support, local companies are compelled to use a significant part of resources in financing the current needs of the working production and - only in rare cases - on development of new technologies in competitive science-intensive civilian production.

2. The structure of innovation-active companies in Russia in the majority of cases is constant; more than 70% of organizations and companies planning in near future for technological innovations in the field of DUT have been carrying them out for the last few years.

3. The deficiency of resources is the main, but not the only factor of innovation activity recession in the area of DUT. The country has started market reforms, where the scientific and research communities proved unable to be ready for work under the new conditions. The scientific and technical developments in the area of DUT do not always become a solid basis for creation of innovative products ready for manufacture and efficient implementation.

4. There are problems of a legal and organizational character in protection and transfer of intellectual property and certification of innovative produce:

a) The lack of a legislative base and a special code for safeguarding innovation and absence of insurance add to the already existing pile of problems. Their rapid introduction as well as the creation of venture funds and innovation infrastructure in general are those necessary conditions, without which it is impossible to ensure innovation breakthrough for both State and private sectors. Alongside large companies, consortia and associations, an important role is played by small companies that provide approximately half of all innovations. According to data at the beginning of 2001, in Russia there were about one million SME's, giving employment to one million persons. But the percentage of SME's which are carrying out innovation activities is only 6%. Nevertheless, these enterprises exist, function and become a more and more important factor of State innovation policy in the area of DUT.

b) Development of a standard legal mechanism covering innovative activities in the field of DUT and the ways of its stimulation have not been fully completed. The systems of institutional transformation, intellectual property protection in the sphere of innovation and mechanisms of introducing it in economic circulation are insufficiently worked out; there are no systems of innovation activity combining support in the area of DUT.

5. For development of DUT it is necessary to develop an innovative process infrastructure, including the system of information support, the system of objective expert examination, the system of certification and encouragement of developments and the system of motivation and stimulation.

6. It is necessary to ensure perfection of a competitive system of selecting innovative projects and programmes for DUT selection. The implementation in civilian branches of the economy of relatively small innovation projects of DUT that quickly pay back, with participation of private investors (including foreign ones) and with the assistance of the State will allow support of the most promising projects and companies, increasing the inflow of private investments, realising the introduction of critical technologies and priority objectives capable of transforming the appropriate branches of the national economy, and ensuring economic progress in the region.

7. Methods of implementing innovation policy in the DUT area are:

- the formation of institutional and legislative conditions for positive changes in the innovation sphere of DUT economy;
- the introduction of intellectual activity results;
- state support and stimulation of investors in putting resources into science-intensive projects;
- high-technology production, eased by introduction of certain tax privileges, State guarantees and credits and the perfection of the tax system with the purpose of creating favourable conditions for conducting innovative activities by all subjects irrespective of patterns of ownership and forms of finance.

8. For effective innovation activities in the DUT area it is necessary to obtain foreign economic aid providing conditions for forming together with foreign partners joint enterprises specializing in science-intensive production and the realization in external markets, taking care of advertising our discoveries and technological inventions abroad, assuring Russian participation in exhibitions and fairs and joining international information systems for data exchange on innovation projects. It is necessary to provide foreign credit lines for developing innovative activities in the area of DUT aimed at fostering interaction in the field with member countries of the EU and preserving and maintaining links with our long time partners in the CIS and other states and ensuring participation of innovation-active companies in international competitions.

9. It is necessary to provide financing of domestic and foreign patenting in the DUT area and support of international patents used in innovation projects. It is expedient to develop the system of competitive selection of projects in the DUT area, including working with international organizations.

10. The important role in expanding the circle of investors funding innovations during the DUT introduction will be given to financial risk insurance. The list of insurance risks will be extended with inclusion in it, in particular, of technical and other risks connected with innovative projects implementation in the DUT area. To reduce risks for private investors, it is necessary to provide for participation of the State in venture funds and appropriate State guarantees in DUT introduction.

The implementation of innovation policy in the field of DUT, taking due account of the potential and interests of all innovation activities subjects and the international, economic, social and political priorities, will promote the reorganization and rise of the national economy of Russia, its successful cooperation with Western countries and its integration into the world market economy.

Cooperation among NATO and Partner SME'S in Concept Experimentation and Advanced Technology Demonstrations

Dr. Velizar SHALAMANOV*, Mrs. Irena MLADENOVA**
*Bulgarian Academy of Sciences - Bulgaria
**Economic Policy Institute - Bulgaria

Abstract. During the first period of restructuring after the end of the Cold War, the armed forces of Partner countries focused on downsizing and utilization of excess equipment. Currently, they face the challenge of modernization in pursuit of interoperability and cooperation not only among the military, but among defence industries as well. Limited experience in new Western technologies among the military and the big defence companies is the reason to look for SME initiatives in NATO and Partner countries. One promising venue is to join the efforts of NATO and Partner SMEs (including academia and, potentially, NGOs) around concept development experimentation (CDE) and advanced technology demonstrations (ATD). These demonstrations serve the purpose of technology evaluation, a test-bed for getting hands-on practice for military and defence industries with modern technology in definition of precise operational and technical requirements and a rapid prototyping approach to the evolutionary acquisition of new weapon systems and the C4ISR systems in particular.

Such cooperation will facilitate the promotion of local capacity and will induce regional cooperation that can lead to joint procurement. We consider this as the best way to establish cooperation with leading Western companies. The experience of Bulgaria and South-Eastern Europe (SEE) is explored as a positive example. The focus is on development of the ATD programme in support of the Force Modernization Plan in Bulgaria, on experience during the last HEMUS 2002 international defence equipment and systems exhibition and participation of Bulgarian SME, academic and NGO institutions in this programme.

This paper will explore the feasibility of creating associations of NATO and Partner SME's, research centres and NGO's within the framework of NATO/EAPC. Comparison will be made with other potential frameworks.

Introduction

11th September, 2001 brought about a new arrangement of the world order. It marks the beginning of a completely new situation where one needs to reassess the very notion of security. International organizations involved in the field of security are supposed to play a decisive role in determining the necessary measures. In this context NATO proves to be the most effective international organization safeguarding stability and security in Europe and has become in practical terms the nucleus of the global coalition against terrorism. The new requirements in this sphere presupposed as well new requirements in the field of the defence industry. New necessities are taking shape for modernization of existing armaments and equipment, for creation of new ones and utilization of others or the destruction of redundant and obsolete articles. This is a necessity both from the point of view of protecting the national sovereignty in the new international environment and of the willingness for more security through a collective system, through international organizations and coalitions.

The participation of Bulgaria in this system, the successful modernization that took place in its military structures and in its armaments and equipment is of paramount importance for the protection of our citizens and for its future role in a dynamic situation. Therefore the reforms in this sector should be seriously and thoroughly considered and implemented, mobilizing the efforts of all State institutions and structures of civilian society.

1. Initiatives

All of the under-mentioned steps could serve as a basis for evaluation of development in the area of modernization of the armed forces and the security sector as a whole, defence industry restructuring, involvement of the academic and NGO sectors and opportunities for international cooperation that can lead to associations of competitive and cooperating SME's in the security area. For many reasons, a success story can be developed in the C4ISR sector with the participation of high-tech SME's, united initially around CDE with ATD and pilot projects in the defence and security sectors at large, with certain regional dimensions, especially in the civilian protection area.

1.1 Round table: Defence Industry and Academic Sector in the Modernization of the System of National Security

With regard to this, the Atlantic Club of Bulgaria, together with the George C. Marshall Association - Bulgaria organized a **round table: Defence Industry and the Academic Sector in the Modernization of the System of National Security**. Representatives of government institutions and experts from the academic community and from the companies of the defence and IT sectors took an active part in the discussions.

All national security and defence planning documents acknowledged the importance of advanced communications and information technologies. Furthermore, the Military Doctrine of the Republic of Bulgaria declared that priority in the modernization should be given to command and control, communications and information systems, intelligence, navigation and surveillance systems (C4ISR).

Significant progress has been made in the last three years in Bulgaria in implementing the decisions for priority development of the C4ISR systems. Reform is needed to modernize C4ISR systems but, on the other hand, the modernization of C4ISR may be an important driver for force modernization, organizational change and the adoption of a new service culture. Of particular importance is the potential of advanced C4ISR systems in enhancing security cooperation and the build-up of cooperative crisis management capabilities in South-Eastern Europe (SEE).

1.2 Round table: C4 Systems and the Invitation to Join NATO in 2002 - Key Projects and Technologies: the Role of Administration, Business and the Academic Sector

In this sense special attention was given to C4ISR at a separate **round table: C4 Systems and the Invitation to Join NATO in 2002 - Key Projects and Technologies: the Role of Administration, Business and the Academic Sector** (February, 2002), organized by the Atlantic Club of Bulgaria with the support of the Ministry of Defence of Bulgaria, the George C. Marshall Association - Bulgaria, the Centre for National Security and Defence Research - Bulgarian Academy of Sciences, and with the participation of representatives from the Presidency, the Parliament, Ministries of Defence, Interior, Transport and Telecommunication, Finance, the State Agency "Civil Protection" and representatives and subcontractors from NATO/EU member countries with projects in the area of security and defence in Bulgaria.

1.3 New association - Business Executives for National Security

The third important dimension of the NGO activity in the area of modernization of the National Security system of Bulgaria is connected with the establishment of a **new association - Business Executives for National Security (BENS)** as an initiative between the Employers' Association of Bulgaria, the George C. Marshall Association - Bulgaria and the Centre for Liberal Strategies - Bulgaria. The new organization whose goal is to facilitate the strengthening of the National Security system by introducing best business practices to government administration and improved Public/Private partnership is really a new dimension – a connection between business (not only the defence sector, but mostly in the civilian sector) and the National Security system. BENS - Bulgaria is already in good cooperation with BENS - US and working to establish links with similar European associations in order to strengthen this important business pillar of National Security.

1.4 Seminar on "Privatisation and Restructuring of Defence Industries, with a Particular Focus on the Progress Achieved in South-Eastern Europe"

A very successful EAPC Seminar on "**Privatisation and Restructuring of Defence Industries, with a Particular Focus on the Progress Achieved in South Eastern Europe**" was held in Sofia (25-26 March 2002), co-organized by the NATO Economics Directorate and the Economic Policy Institute of Bulgaria. The seminar followed the Economic Policy Institute's commitment to contribute to the discussions on economic aspects of Bulgaria's future membership in NATO, and built on the results of an international conference "**Restructuring of the Defence Industrial and Technological Base in New and Prospective NATO Members**" EPI organized in June 2000 (Sofia).

The EAPC Seminar gathered more than 80 participants (high-level government officials, representatives of defence companies, including leading multinationals, academics and independent experts) from 18 countries (NATO members, applicant countries and international organizations). The main issues discussed focused on:

- Challenges to Defence Industry Restructuring and Opportunities for Cooperation;
- Adaptation of Defence Industries to the Requirements on Material Standardization, Quality Assurance and Interoperability;

- Modernization and Competitiveness of Defence Industries; New Investment and Management Strategies; Communication and Information Technologies; Dealing with International Consortiums; Projects on Development and Manufacture of Defence Products;
- The Role of Ministries of Defence in Defence Industry Privatisation and Restructuring; Implications and Specifics of Future Relationships.

1.5 1st International Conference – Defence Research and Technologies "Modernizing the Forces – Cornerstone of the Bulgarian Way to NATO"

The 1st International Conference – Defence Research and Technologies "**Modernizing the Forces – Cornerstone of the Bulgarian Way to NATO**", under the patronage of Nikolai Svinarov, Minister of Defence and Academician Ivan Iuhnovski, President of the Bulgarian Academy of Sciences, was held within the 5th International Defence Equipment Exhibition "HEMUS 2002" (30th May-1st June 2002) at the International Fair in Plovdiv. It focused on C4ISR, coping with new risks to security and accompanied an Advanced Technology Demonstration of solutions based on advanced IT. The two sessions focused on:

- C4 Systems in Support of the Bulgarian Forces Modernization (System Concepts and Force Development; Advanced Technology Demonstration; Implementation of COTS Technologies; International Cooperation in C4ISR System Development) and
- Research and Development of Armaments in Support of the Modernization of the Bulgarian Forces (System Concepts and Operational Research; Project Planning, Management and Evaluation; International Co-operation).

The exhibition, conference and ATD primarily focused on the long-term Modernization Plan of Bulgarian armed forces and the new challenges to security and regional cooperation.

2. Modernization of the Armed Forces

In the process of elaborating the Concept of National Security, the Military Doctrine, Plan 2004 for Modernization of the Army, and the Action Plan for Membership, a less discussed theme is the modernization of the system of security as well as the issues related to the defence industry of Bulgaria. Therefore, in the programme of the new government, although a fifth priority, it is written that: "… carrying out modernization and rearmament of army and forces with armaments, machinery and equipment that will correspond to the missions and tasks, implemented in a dynamically changing and risky environment".

This will lay the foundation work for developing a plan for modernization of the armed forces. The process should reach maximum expansion, not confined only to the Ministry of Defence, but embracing all the elements of the system of national security, especially the Ministry of the Interior and Civil Protection Agency. The Bulgarian defence industry, the academic sector and non-governmental organizations should be given a more visible role in preparation and implementation of the plan for modernization.

A new system for development of armaments will create conditions for concentrating all the efforts for impartial spending of resources allocated for the implementation of the armaments plans in the hands of the Ministry of Defence. This system should account as well for the presence and capabilities of the defence industry. TEREM SHC stands most closely to the Bulgarian Armed Forces and is best adapted to their particular needs. Cooperation among the rest of the defence companies and TEREM SHC is the most direct path to the programme of force modernization and programmes for utilization. The latter are a matter of priority and easier to finance, including the use of foreign funds.

The US Atlantic Council Report on strategic opportunities for the Bulgarian defence industry in transformation, reorientation and integration into NATO is a step forward in the process of analysis and the search for positive options of development. One of the options for our country is to use the means of the Partnership for East Trust Fund. The latter provides opportunities for companies from the countries participating in the Partnership for Peace, which can develop and produce defensive items under the standards of NATO. The companies concerned are going through a special procedure of certification financed by the Fund. The Bulgarian State is responsible for laying down prerequisites for certification of production and classification of products in compliance with the contemporary industrial and military standards. Certification will open prospects as well for more active participation of Bulgarian companies in different offset programmes and subsidiary contracts, even when Western companies are realizing the transactions.

The Bulgarian defence industry can be modernized in two principal directions; through foreign technologies and investments or on the basis of its own national resources. In both cases the academic sector has not yet realized the role which it could play for the defence industry.

A number of institutes related to the big defence companies have been closed down. The same goes for institutes related to the Ministry of Defence. However, we have a national institution like the Bulgarian Academy of Sciences (BAS) and several powerful technical universities, which possess capacities for resolving the key issues; transition from Russian to Western technologies and adaptation to Western technologies. A new technological breakthrough is possible toward traditional markets, not, however by exported classical weapons, but through modernization of Russian armaments by Western technologies.

Thanks to the cooperation with BAS last year, the Ministry of Defence made an important step towards modernization at a national level. The area for scientific research and development in the interest of defence has been structured. The BAS covers 80% of this structure. A database with scientific and research potential is being developed. It is necessary that it is expanded with the potential of universities, higher educational establishments and some High Technology firms from the defence industry which maintain research units. This is the way to explore the areas where Bulgaria can seek its competitive edge.

The resources of the BAS, where 60% of the scientific potential of the country is concentrated, have not been used properly by all agencies related to national security. The most concerned about the appropriate direction of the modernization of the armed forces are the Bulgarian defence companies, which are producing military equipment with a military purpose and special items in real terms. It is more natural to finance innovation projects of the academic sector and to provide opportunities for prominent Bulgarian scholars and researchers to work for the national defence industry. Together with non-governmental organizations, the academic community could participate more actively in initiatives for training managers and specialists from the defence sector on issues of European and Euro-Atlantic integration, including language qualification and models of management.

Weak knowledge of foreign languages, especially at the medium managerial level, is a serious handicap in finding Western partners for our defence industry. The State has taken care of the restructuring of the sector so far, but has failed to attract any strategic investor there. The reason is the lack of a working promotion strategy for our defence industry and the weak coordination among different departments. Not enough efforts have been exerted to attract the attention of foreign military and trade attachés stationed in Bulgaria. Insufficiently used is the potential of our Embassies and the Bulgarian lobby around the world for generating commercial, industrial and investment interest in the respective countries towards Bulgarian military enterprises.

A serious analysis of the place of the defence sector in the field of modernization of the armed forces, consolidated by the defence companies and ministries themselves that are related to these problems, but realized by the academic and non-governmental sector, will bring about visible progress. During the round table discussions on force modernization many serious conclusions were made:

1. **There is no unified plan for modernization** of the system of national security. The technological gap is significant, the system for development of armaments, machinery and special systems **does not match the level** required for membership in NATO and the European Union.
2. **There is poor coordination** among the Ministry of Defence, the Ministry of the Interior, the Civil Defence Agency and the Ministry of Economy on the processes of modernization. The Ministry of Defence is working on a research programme for modernization but it does not provide for incorporation of the Ministry of the Interior and the Civil Defence Agency in the process, so that a unified plan for modernization of the system of national security can be worked out. This is one of the most important conclusions made after the "September 11th attack", especially in the field of management systems.
3. The companies in the defence sector **are ill-informed on the government's plans** for modernization of the system of national security. They have little knowledge of the prospects for cooperation with strategic partners from member countries of NATO and the EU in the process of modernization, utilization and rearmament.
4. **The link** between the defence industry and the academic sector **has been disrupted** and there are practically no serious projects for the introduction of new technologies. **There is no effective standard base** for concluding agreements for scientific research, developments of prototypes, demonstration and introduction of new technologies with participation of the academic sector and Western partners.
5. There is a lack of a **system for training of managers and specialists** in the defence sector, adequate enough to prepare specialists for the Ministry of Defence, the Ministry of the Interior, the Ministry of Foreign Affairs, the Ministry of Economy and the Civil Defence Agency, on issues of European and Atlantic integration, including language qualification and models of management.
6. **Opportunities for implementation of** offset programmes **remain underdeveloped. Defence and academic sectors do not offer in a sufficiently active manner any solutions for modernization to the competent departments.** The process of privatisation in the defence sector has failed to attract any serious Western partner. **It did not bring about innovation in the sector neither did it introduce any new know-how, either technological or managerial.**
7. **In fact, there is no dialogue** in a broader format – between Parliament, the President's Administration, the Government, the Ministry of Defence, the Ministry of the Interior, the Ministry of Economy, the Ministry of Foreign Affairs, the BAS, the National Fire Department and Western partners - on the basis of which is the sole possibility of defining an environment for effective modernization. There is neither a "List" available with the main projects for modernization, rearmament and utilization nor an effective mechanism in

the Public Orders Act allowing long-term management of major projects beyond the framework of one department.

As a result of discussions, a number of proposals were defined and it is important to mention that already, step-by-step, some of them have been implemented or are in the process of implementation:

1. Inclusion of the national defence industry, the academic sector and non-governmental organizations in the analysis of the status of the system of national security and **working out a plan for modernization** with the participation of non-governmental organizations, the defence industry and the academic sector.
2. **Hearings of the Ministry of Defence, the Ministry of the Interior, the Civil Protection Agency, the Ministry of Economy and the Ministry of Foreign Affairs**, on security and defence issues regarding:
   - **The pace of modernization of the system of national security** and preparation of a complex plan up to 2015 for comprehensive rearmament.
   - The **state and capacities of the defence industry and its participation in the modernization of the system of national security** and preparation of a complex plan up to 2015 for comprehensive rearmament.
   - The state and capacities of the academic sector and its participation in the modernization of the system of national security and preparation of a complex plan up to 2015 for comprehensive rearmament.
3. Preparation and conduct of a **series of round tables and conferences** on modernization of the system of national security in parallel with the HEMUS 2002.
4. Preparation of a **Project for a National Research Centre in Security and Defence,** outside the structures of the Ministry of Defence, the Ministry of the Interior and the Civil Defence Agency, but under the control of Government and Parliament. The Centre is supposed to work with the ministries, the national defence industry and foreign partners for modernization of armaments and equipment. It should be financed by the budget, by a percentage of the turnover of the trade in weapons and other special production, by contracts with customers of research results, by transfer of technologies, by training of specialists and other scientific and technical services (as a part of the Modernization Plan).
5. Working out of a **National programme** for:
   - Restructuring of state-owned companies of the defence sector;
   - Training of managerial and support staff;
   - A scheme for privatisation with participation of strategic partners – member countries of NATO, with special attention paid to TEREM JSC and Electron Progress (a section of TEREM and its privatisation "in parts" without which a strategic Western investor would not be allowed to participate).

3. Challenges to the Defence sector in Bulgaria

The issues and opportunities faced by the Bulgarian defence industry have been of particular interest to both experts and society since the beginning of transition. The defence industry was a tricky issue to discuss, partially due to the social impact from the restructuring process (about ¾ of the workforce in the defence industry was laid off during the restructuring process), as well as the vested interests, both national and foreign. This does not make it differ much from other sectors of the national economy that are going through restructuring and privatisation.

But when dwelling upon future strategies that could revive the Bulgarian defence industry, one should bear in mind a number of factors. These factors are characteristic to the defence sector alone, and give a more realistic picture of global trends:

3.1 Warfare is a different art nowadays

The end of the Cold War era brought about new challenges to defence industries. Making war no longer involves massed land and naval forces, equipped with myriads of conventional arms to survive through prolonged battles. Rather, conflicts tend to be very short, with minimal casualties, applying information-based, technologically sophisticated weapons and systems (this is the case with fighting terrorism as well – terrorists usually apply low-level technologies). Moreover, communication and information are keys to winning-war strategies.

3.2 The Defence industry has changed globally

- Both in the US and in Europe there is pressure for consolidation. And much of this has already taken place. Presently, prime contractors on both sides of the Atlantic are immense corporations and they make up a rather short list.
- The defence industry environment is rather different from commercial industry and does not follow well-known market principles. First of all, usually the major buyer of defence production is the government, and in cases of exporting – the governments of other countries. Additionally, the defence industry is seen as part of the national security system of a country. That is why defence companies are closely scrutinized and subject to strict regulation.
- Due to changed concepts of conflicts, defence companies have to shift to production of high-tech sophisticated weapons and systems. The demand for conventional arms is steadily decreasing. For the development of new arms complying with the new requirements, research and development is crucial, and R&D expenditures should be substantial in order to ensure it. Partnership is the other key concept for prosperity. Prime contractors need the smaller sub-contractors in order to provide high-quality services and smaller companies cannot obtain large contracts on their own.

Such a review of the global environment and trends is crucial for identifying the competitive advantage(s) that provide long-term opportunities for the development of Bulgarian defence companies.

The next stage of that review focuses on the factors that shaped the current status of the Bulgarian defence industry. To a certain extent, these factors could be derived from the global context, and the ability/inability of Bulgarian companies to adapt and to define and exploit the opportunities:

- Since the beginning of a systematic transformation in Bulgaria, the defence sector has gone through a painful and prolonged process of reform. Defence sector output drastically fell; defence companies lost their privileges, and were exposed to a greater extent to the market economy mechanisms and the challenges of shrinking export markets.

- Bulgarian defence companies failed to attract strategic investors. Privatised by MBO's, they missed, or at least postponed, the chances for modernization. Bulgarian companies could not obtain the modern technologies and equipment, the transferred know-how, the financial support and the management skills needed to take them to the next level in the restructuring process. As a result, Bulgarian companies are losing the competitive edge they once had and the foreign markets they used to dominate.
- Bulgarian defence companies, like many commercial companies in other sectors of the Bulgarian economy, are lacking the strategic planning that could set the goals and steps, define the resources needed, and mobilize the personnel for their achievement. However, many companies managed to diversify and undertook production of civilian and dual-use articles. And these companies are much better off. Many others have still to re-think their prospects and seek to identify the appropriate market niches in order to become sustainable.
- NATO membership opens up new opportunities and poses new challenges to Bulgarian defence companies. It should ensure better prospects for cooperation of Bulgarian with NATO members' defence companies. It will bring new options for participation in joint R&D projects, procurement, as well as subcontracting. On the other hand, NATO enlargement will in fact increase the market for the leading multinationals. In order to cope with this pressure, Bulgarian defence companies will have even more incentives to seek opportunities for subcontracting and joint ventures.

4. The role of small companies in the defence industry

The challenge of the defence industry restructuring is to make the defence companies viable entities, increasing their efficiency and competitiveness. Although the restructuring process in Bulgaria started somewhat later compared to other CEE countries, at present the bulk of the defence companies is privatised (through MBOs), and about 20% are entirely in private hands.

Small companies in the defence industry, although differing to a greater extent to their commercial counterparts, share many of the problems:

- **access to financing**. Although possessing significant amounts of fixed assets and production capacity (unfortunately, often obsolete), they lack access to financing.

- **access to markets**. Their domestic demand has decreased substantially during the transformation and they have lost many of their traditional export markets.

- **access to new technologies**. Small companies do not have enough financial resources to invest in R&D.

- **adoption of new management and marketing approaches**. Small companies, again usually due to financial reasons, but sometimes due to poor strategic planning, do not invest in training, management and marketing.

- **finding the niche**. It is critical for small companies to analyse the market and find the appropriate market niche that could give them sustainable competitive advantage.

These difficulties could be overcome if a company analyses in-depth its activities and designs an adequate strategy. For commercial sectors of the economy, SME's usually have a number of scenarios for development, depending on different combinations of the factors and characteristics of the product they offer, the industry they belong to, the market they focus on, and the resources they have access to. In general, applying this approach to the defence industry could give some economic insights and support strategic planning.

Market analysis should keep in mind the unique characteristics of the defence industry. Supply and demand in the defence industry are quite different from supply and demand in commercial sectors. The buyer of defence goods (demand) is the government. Moreover, the defence companies could hardly "re-focus" and find another "target group" for their product. Additionally, the market is constrained by considerations of national security. With the case of exporting, the buyer is again a government, but of another country (directly or indirectly). The nature of demand sets some requirements and specifics to defence industry producers. Besides, not all markets are open to all players (for example the US "buy American" act for defence products).

On the supply side, small companies are rarely prime contractors for big projects. But they could be very helpful in providing certain services or elements to the prime contractor (sub-contracting). Huge investments and R&D expenditures are needed in order to keep up with the industry development (barriers to entry) and could keep away smaller companies. Expanding technology requires intensive investment and expensive R&D activities, and individual companies, especially small ones, have to seek cooperation with strategic partners to withstand the pressure.

Possible sustainable strategies for small companies, following such an analysis, include subcontracting, establishing joint ventures, finding a niche. And the more specific options for Bulgarian companies vary from making it on their own (which could hardly be sustainable), establishing partnerships and cooperation with the EU (and becoming part of the European defence basis, participating in European projects) and establishing partnerships and cooperation with US industry.

The government is in a good position to assist SME's on their way to sustainability and profitability. A very good sign in this direction is the large number of intergovernmental bilateral cooperation agreements Bulgaria has – with Germany, Italy, France, Russia, Ukraine, etc. Bulgaria is an associate partner in the Western European Armaments Group (WEAG), and as such could participate in specific armaments projects (taking part on the same basis as full members, including the contribution of an appropriate financial share).

Although providing a good framework at government level, these agreements have to be backed up by a sufficient number of company-to-company agreements for cooperation and specific contracts. Bulgarian companies are not always proactive enough to get much use of such opportunities.

Small companies need to focus on the development and strengthening of their core competencies. In many cases this would involve a reshaping of the product portfolio, and getting rid of unprofitable and uncompetitive articles. Reorganization would involve downsizing (employees, obsolete equipment and technologies) and developing "profit centres".

Defence companies should build up their strategies in order to make more efficient use of resources. Recent trends of reduction in defence budgets and the increasing pace of technological progress justify this need. Besides, more and more defence companies will have to compete on the labour market with commercial sectors in the country and abroad for attracting and retaining well-educated and highly skilled experts.

Smaller companies are even more subject to the pressure to be efficient and cost-effective, as they are less government-protected (large company bankruptcy would cause immense negative effects on a country's economy and could lead to major social disturbances - defence companies are also seen as part of the national security system – all this makes the bankruptcy of a large defence company undesirable and almost improbable). The answer for SME's is restructuring and downsizing, cooperation, mergers and joint ventures.

Cooperation has many faces. SME's should exploit as many of them as possible:

Cooperation with the academic and NGO sectors. This offers wide opportunities for SME's, and is almost imperative, considering their own poor R&D basis. Many of the research institutes associated with Bulgarian defence companies and the Ministry of Defence have been closed down. The remainder is not capable of performing high-quality R&D activities mainly due to financial restrictions. The Bulgarian Academy of Science, the technical universities, as well as a number of NGO's specializing in security and defence issues, supply qualified motivated experts and could be helpful partners.

Cooperation with large defence companies/multinationals. This cooperation could take the role of participation in joint projects, subcontracting, etc. Cooperation with large defence companies is crucial for the transfer of advanced technologies and know-how, progressing to bigger markets, and adopting modern strategies and approaches to management, marketing and human resources management.

Cooperation with small companies. Cooperation within the segment also provides good opportunities and should not be neglected. Together, small companies could qualify for bigger contracts, could initiate/participate in bigger projects.

Balancing cooperation and competition would remain a key challenge, as well as adapting to the new order and the economic environment and potential threats. One very promising area for the Bulgarian defence industry is the development of C4ISR projects. It also provides a good illustration of the joint efforts of companies, government and academic and NGO sectors for modernization.

We believe this area can be used as a national pilot programme, as C4 is critical for interoperability, for new technologies' insertion in National Security structures, for cooperation between administration and business and transfer of "implementation know-how" between the security sector and other sectors of State administration.

An NGO project on Bulgarian Security and Defence Information Network, using new technologies and cooperation among the Administration, business, academic and non-governmental sectors- "**Bulgarian Action Plan to Advance C4ISR Infrastructure**" should be accomplished before Hemus 2002 and distributed among participants and other interested organizations from the Administration.

In practice, with the number of initiatives, round tables and follow-up activities, the NGO sector set up a good ground for the Administration to prepare the Force Modernization Plan in close cooperation with the academic sector and business, attracting interest from the Parliament, media and political parties as well as our foreign partners. That is serious achievement.

5. Key role of C4ISR projects

After the most difficult political decision was made, the next single most important factor for security and defence reform was *modernization* of the armed forces and the security sector as a whole. In this effort, C4ISR projects will be of crucial importance. Their implementation will have a multidimensional impact:

- A number of important and expensive projects have already been launched;
- There are many positive developments in the standard and expert environment;

- Implementation of these projects will lead to reform processes in other areas;
- C4ISR will provide interoperability and security and will integrate other structures and processes together in one security and defence system of a new type that can be considered as the *e-Security and Defence* part of the e-Government.

Modernization as a whole includes *utilization* of the surplus of obsolete equipment that is difficult to support, not interoperable, or simply not adequate to the requirements of the Military Doctrine. The second component is *modernization* of equipment that can serve for the next five to ten years. The third component is the *acquisition* of the new, probably in some cases second-hand, equipment that will serve for at least the next twenty years.

Of particular importance is the harmonization of modernization plans with risks and threats, a complete set of missions, the structure of the armed and security sector forces, a system of training and life-cycle support capabilities of the country and, last but not least, available resources, R&D and industrial capacity.

Additionally, IT can dramatically improve the quality of implementation of the modernization plan through a system of programme management and overall life-cycle management. [3] It allows involvement in modernization planning of the administrative organization of the security and defence forces, industry, the academic sector, non-governmental organizations (NGO's) and the broader civilian society.

C4ISR systems are specific, because they are based on the most rapidly changing technologies. Interoperability and dual-use issues are of critical importance for them and the life-cycle is based more on evolutionary prototyping and development than on an "under key" approach of developing systems which can be operated for the next few years without significant upgrade and improvement. [4] These systems are very sensitive from the security point of view because, first of all, they keep all the information and, secondly, they control all other systems.

Furthermore, C4ISR systems and related projects have a crucial role because they integrate existing platforms in combat complexes and provide interoperability in the key C2 areas dealing with the most important resources - information and knowledge. At the same time exactly, these systems are becoming obsolete very rapidly and, as was mentioned before, this impacts on information security which is critical under rapid technological changes and convergence with civilian systems.

Finally, C4ISR systems are unique because of the integrating role they play in combat complexes and the procedures for life-cycle support. In the context of the process of rethinking which is underway, and the changes after the tragedy of September 11th, 2001, it is clear that C4ISR systems will be even more important for, and, at the same time, strongly influenced by the coming strategic shift.

While debating modernization issues, the following aspects of the role of C4ISR projects have to be covered:

- Life-cycle management of C4ISR systems;
- Business aspects of C4ISR system development, especially integration and cooperation among companies, both suppliers and system integrators;
- Modernization at large and the role of the Administration, business, the academic sector and NGO's;
- Dealing with international consortiums, new investment and management strategies.

6. C4ISR as an element of the security sector modernization and life-cycle management

Currently, the Bulgarian Ministry of Defence, jointly with a US contractor. accomplished a comprehensive force modernization study (FMS). The results of the study will be consulted by NATO through the Defence Support Division of the Alliance and will serve as a basis for development of a long-term plan for modernization of the Bulgarian armed forces. Ideally, the plan will provide for coordinated development of platforms, the defence industry and the academic sector. In the long term, the plan should outline the development of aircraft, surface-to-air missile squadrons, ships, main battle tank modernization, armoured personnel carrier modernization and artillery, including modernization of anti-tank weapons.

Another aspect is the planning of the utilization of the large inventory of platforms, small arms and light weapons, and ammunition. Platforms will be assessed in the context of the doctrines and emerging C4ISR systems, which are further ahead in modernization than any other element of the military complex. The main criterion should be how efficient they will be in exploiting the comparative advantages of Bulgaria. Logically, the companies involved in C4ISR projects that have capabilities also in the area of platforms can be a factor for integration in the overall modernization effort. On the Bulgarian side, such a potential factor is the company TEREM because of the scope of its competencies and experience in dealing with the MOD and other elements of the security establishment. On the other hand, TEREM is still State-owned, so it may potentially serve as a kernel for building a "National Defence Industry Consortium" - a private company (SHC), that can be flexible in international cooperation and, eventually, allow for foreign investment through privatisation.

The study covered nine different areas of modernization; of weapon systems and equipment of the Army, the Air Force, the Navy, Logistics, C4, ISR, Defence and dual-use infrastructure, Utilization and the defence industry.

Integration of the platforms through C4 is essential. At the same time, C4 infrastructure needs its own integration through a control centre (Network Control Centre (NCC)). The development of NCC is part of the implementation of C4 projects. Integration is only one of the requirements. Rather more difficult to achieve is interoperability with NATO, flexibility, security, readiness, etc.

Finally, the C4ISR system is planned on at least three levels: as a set of equipment with certain capabilities and fixed structure; as a task/operation-oriented temporary and mostly mobile structure; and as a real-time response for immediate event-driven reconfiguration. Normally, this planning is directly related to operational planning and planning of the use of platforms that will be controlled through the C4ISR system. [5] That turns *the environment for configuration planning and management* into an essential factor for the success of any C4ISR system.

7. Organizational issues in the C4ISR area

In order to address more efficiently the issues of C4ISR life cycle management, the Bulgarian Ministry of Defence established the institution of a Chief Information Officer (CIO). Further, it introduced the *Manual for Life-Cycle Management* that clarifies organizational structures and procedures in order to maintain such a system from the initial idea to the system's disposal and utilization. From a national point of view, after regulation of the issue within the MOD, it is important to extend these efforts outside the MOD and to prepare at national level the environment for management of the life cycle of C4ISR systems. That would include introduction of CIO institutions, a Programme Office for C4ISR, and designated operators of the C4ISR in other government agencies.

On the other hand, there is a need for cooperation among the companies that are already involved in modernizing the C4ISR system. Thus, American and European companies may coordinate their efforts and cooperate more closely with the Bulgarian defence and civilian industries, for example in offset programmes, E&T, R&D, test-beds (research, demonstration and certification centres), etc. TEREM can cover all types of platforms and some kinds of C4ISR equipment, but there are many others that can join the club. As discussed previously, the "club" has to be gathered around a private company with the participation of the MOD and the Bulgarian Academy of Sciences.

Such an organization would provide for cooperation at many levels:

- overall management of the process and alliance to modernize the defence and security establishment;
- command and control centres (including NCC);
- C4ISR architecture;
- platforms.

The issue of defence industry involvement is closely related to the restructuring and the strategic plan for the development of this sector. A recent study by the US Atlantic Council working group was a positive first step [6]; it has to be deepened and converted into a full-scale strategy and implementation plan.

When such an environment is established, projects can be divided among the companies through partnership agreements, while still allowing for transparent competition and cooperation.

Non-governmental organizations (NGO's) can also be instrumental in providing transparency and coordination. Among many NGO's that are potential participants in such an arrangement, we would refer to the "George C. Marshall" Association, covering broadly the security area from security policy to security sector reform, and then to acquisition and defence industry restructuring. Given improved coordination, the activity of AFCEA (Armed Forces Communications and Electronics Association) in Bulgaria can give a substantial boost to inter-agency and international coordination. Another opportunity is the recently established Bulgarian "Business Executives for National Security" (BENS), an organization similar to the US "Business Executives for National Security", and of course the Atlantic Club of Bulgaria is a prominent NGO with great influence in these areas.

The potential for regional cooperation has not yet been utilized. The regional security cooperation in South-Eastern Europe, involving Albania, Bulgaria, Croatia, Macedonia, Slovenia, Romania, Greece, Turkey, Italy and, hopefully soon, Yugoslavia and Bosnia-Herzegovina, has a significant potential. Then, it would be possible to create a network of networks in which the region of SEE cooperates with Central Europe, the Baltic, the Caucasus, Central Asia and the Mediterranean region.

The SEE Defence Ministerial (SEDM) process provides a feasible framework for cooperation and *joint procurement, joint training and certification.* A priority area for joint procurement will be C4ISR. With this in mind, SEDM countries initiated the establishment of a Regional Coordination Group (Committee). We envision further progress with US support; one opportunity for discussion should be provided at the international C4 Conference in May 2002 during the HEMUS 2002 Defence Exhibition. The main topic of the conference will be life-cycle support of C4ISR systems on a national, regional and interregional basis. We expect participation of major American and European C4ISR contractors.

It is clear that only good cooperation and harmonization of interests and capabilities can generate such a project. At the same time, this project is not starting out in a void - there are many accomplishments in the area of C4ISR and a number of ongoing projects. Therefore, right now, consolidation of all existing, ongoing and planned projects within the large Information Grid (IG) programme is very important and will be a good example for the even larger e-Government programme. But to start with, the National Security System is essential especially because of Bulgarian integration into NATO and the EU, membership in the UN Security Council, future chairmanship of OSCE, and the growing demand for security cooperation after September 11th, 2001. The IG will integrate all elements of the National Security System efficiently and will provide international integration, as well as higher interoperability with NATO/EU countries.

8. Dealing with international consortiums, new investment and management strategies

When analysing resources needed for recapitalization of the armed forces, it is clear that there is a need for a really radical and non-traditional approach. Such an approach can result from close cooperation among the Administration, business and academic sectors, supported by NGO's and dealing with strategic partners from NATO countries. It is why BENS-Bulgaria (Business Executives for National Security) was established as a non-profit non-governmental organization whose goal is to facilitate the strengthening of the National Security System by introducing best business practices in government administration and improved Public/Private partnership.

BENS will focus on these major project areas:

- ***Action plan for recapitalisation of the Bulgarian Security Sector – mostly the Bulgarian Armed Forces:*** this is an important area that addresses the utilization of extra or obsolete equipment and infrastructure; the issue of restructuring current, top-heavy personnel structures; the vision of how to modernize and build new infrastructure; how best to modernize and acquire new equipment; how to meet the requirement of recruiting and retaining skilled people. The goal in this area is to motivate the Government Administration to prepare a comprehensive Force Structure Plan, a Modernization Plan and a Professional Resources Plan. In addition, the Ministry of Defence should request an Armed Forces Recapitalization loan from the US and the EU. The request should be in the form of Foreign Military Funds for equipment and services. The rationale should be transformation of a part of the Bulgarian debt based on the formula "debt for common security", thus linking large-scale procurement programmes with direct and indirect offsets in strong and important areas of our national economy.

Plans will be developed and implemented to link business activity and interests with the recapitalization of the Armed Forces - modernization and professionalism for increased participation in multinational operations - up to 600-700 people, not only NBC and engineers, but also SOF, helicopter units, MEDEVAC, as well as an additional 500 personnel from police/gendarmerie.

- ***Action plan for advanced C4ISR infrastructure in Bulgaria, SEE and Black Sea Region-***
- A key element of the programme will be C4ISR national infrastructure – with the Bulgarian Information Grid as the driving force, capabilities multiplier, enabler and amplifier of our geo-strategic and geo-political comparative advantages and system organizer for all other elements of the security sector.

- ***Action plan for public/private partnership in security and defence – outsourcing and implementation of successful business practices in security and defence management***
  - A very important business case for BENS-Bulgaria is to find ways to increase investment in Bulgaria that will modernize our industry with "state-of-the-art know-how" and a plan to reduce part of our foreign debt as addressed in this business case.

This comprehensive programme should be a joint effort with the private sector, in order to assure that this government cost will not simply increase our debt, but will reduce it dramatically in the mid-term period. This will result from positive economic growth, new employment, new business enterprises, modernization of defence and commercial industries, application of new technologies in all areas of the enterprise at all levels of Government in both defence and national security. Other elements will cover air defence, the air fleet, the sea fleet, land forces (especially Special Operation Forces) and of course other elements of the security sector in the Civil Protection Agency and MOI.

The package investments in the key infrastructure (transport, pipes, communications, energy, tourism, health and human services, etc.) will, of course, also include restructuring of existing State-owned defence companies and other important privately-held corporations.

Creating associations of NATO and Partner SME's, research centres and NGO's within the framework of NATO/EAPC is a powerful tool for addressing modernization issues and improving economic cooperation. In comparison with other potential frameworks – regional, or around certain large modernization/re-equipment programmes, the NATO/EAPC is probably the largest platform to be used in facilitating all potential programmes. Certainly all the programmes have to start with CDE through a series of ATD, prepared and conducted by associations of high-tech SME's, based on good academic support.

New management strategies should pay close attention to the strengths of the Bulgarian defence sector, which includes highly qualified specialists whose potential should be utilized appropriately. Additionally, Bulgarian defence companies will need effective approaches in order to attract and motivate more experts in the future. They will have to cope with competition on the labour market as well.

9. Conclusions

The modernization of the National Security System is probably the single most important task of the current stage (2002-2010). Of course, it is dependent on the reform of the security sector, the process of professionalization of the Armed Forces, the increasing participation in international operations and the overall modernization of the country. At the same time, modernization of the National Security system is a national project and public investment can motivate many other positive changes in society and the State. That is why the role of NGO's is so crucial - to prepare the ground and to support difficult decisions, to provide transparency and to protect interests of society in the long term (longer than one Government mandate) and to motivate best use of national academic potential in the area of R&D, E&T and decision-making support.

NGO's and the academic community in Bulgaria have a very good potential to cooperate with the government as well as with the SME's in the field of security issues in general and, in particular, the development of advanced technologies. Moreover, there exists an appropriate basis for this. More opportunities should be identified and explored for wide-range cooperation among Bulgarian SME's and foreign partners, governments and academia for mutual benefit. Bulgarian SME's have the potential to turn into profitable sustainable units if adequate strategies and approaches to overall management are applied and modern technologies and know-how are imported. SME's should find a niche and concentrate on their core competencies. In order for all this to happen they will need to closely cooperate with local and foreign partners. Joint projects and forums should facilitate the transfer of expertise of defence industries in NATO member countries.

The role of C4ISR projects is essential for overall modernization efforts, because of the influence on force structures and decision-making processes. C4ISR developments impact security, interoperability, horizontal and vertical integration, cooperation with the civilian sector and the use of civilian resources for security and defence. They are unique from a technological point of view and influence life-cycle management.

On the Government side, certain arrangements have already been made in the area of requirement definition, planning and programming, acquisition, R&D, E&T, testing and evaluation. Partnership among business companies and between business companies and Government is essential. There is an opportunity and a need of additional efforts in this area to provide a better environment for successful implementation of C4ISR projects. It can be a driving force to form alliances, to introduce new strategies for investment, offset programmes and effective management of the modernization of the country at large.

Analysis of the three main aspects of NGO activity in this field in the last three months proves that it provided the establishment of the best possible environment for development and implementation of the Force Modernization Plan, definition of the C4ISR programme as a pilot effort in this area and the establishment of a link between modernization of the National Security sector and modernization of the country as a whole through the activities of BENS. It is a good start for future successful cooperation between the Administration, NGO's, the academic sector and business in other areas of public interest.

References

1. Velizar Shalamanov and Todor Tagarev, *Re-engineering the Defence Planning in Bulgaria* Research Report 8 (Sofia: Institute for International and Security Studies, December 1998), 28 pp. Available at http://www.isn.ethz.ch/isis
2. Ralitza Mateeva, Todor Tagarev, Tilcho Ivanov, *Economic Benefits from Bulgarian Membership in NATO*, ed. Velizar Shalamanov (Sofia: George C. Marshall-Bulgaria with the Centre for Liberal Strategies, October 2001), - 69 pp.
3. *Defence Industry and Academic Sector in National Security System Modernization*, Round Table Discussion with Report #3, ed. Velizar Shalamanov (Sofia: George C. Marshall-Bulgaria with the Atlantic Club in Bulgaria, 30 October 2001).
4. Velizar Shalamanov, C4ISR infrastructure - armed forces and industry modernization tool, EAPC seminar on defence industries in SEE - Sofia, March, 25-26
5. *Bulgarian Action Plan for Implementation of Advanced Communications and Information Technologies in the Security Sector*, Round Table Discussion with Report #5, ed. Velizar Shalamanov, Todor Tagarev (Sofia: George C. Marshall-Bulgaria with the Atlantic Club in Bulgaria, 05 February 2001).
6. Velizar Shalamanov, Ralitza Mateeva, "Readiness Assessment and the Economic Benefits of NATO Enlargement", PfP Consortium WG on Future of NATO presentation, March'02, Bucharest

7. *Bulgaria in NATO – 2002*, Institute of Regional and International Studies, March 2002.
8. *Analysis of Bulgarian readiness to join NATO in comparison with other MAP countries and new NATO members*, Project under contract with Bulgarian Parliament with other NGO, ed. Lubomir Ivanov and Velizar Shalamanov, February 2002

Notes

1. *Military Doctrine,* Article 97.
2. Command, control, communications, computers, intelligence, reconnaissance and surveillance [systems].
3. Todor Tagarev, "Economic aspects of defence modernization," in *Economic Benefits for Bulgaria from Joining NATO* (Sofia: Centre for Liberal Strategies, 2001), pp. 25-42.
4. *C4I Study for Bulgaria: Final Report* (USAF ESC/MITRE, January 2000).
5. Velizar Shalamanov, *Life Cycle Support of Information Systems Management*, Doctoral Dissertation (Kiev: Air Defence Radioelectronics Institute/Institute of Cybernetics, March 1991). in Russian
6. Curtis M. Coward and Jeffrey B. Bialos, *The Bulgarian Defence Industry: Strategic Options for Transformation, Reorientation and NATO Integration* (Washington: The Atlantic Council of the United States, July 2001).
7. See for example Velizar Shalamanov, "Concept and Problems for the Development of National Communications and Information Processing Infrastructure," in *Proceedings of AFCEA-Europe Warsaw Seminar* (Warsaw: AFCEA Europe, 1993), pp.11-13.

VELIZAR SHALAMANOV is advisor to the President of the Bulgarian Academy of Sciences on national security and defence issues and Chairman of the "George C. Marshall Association" – Bulgaria. Based on his experience as former Deputy Minister of Defence (defence policy and planning, November 1998 - July 2001) he serves as an advisor to the Parliamentarian Committee on Foreign Policy, Defence and Security and as Director-Strategic Studies at the Atlantic Club of Bulgaria. Dr. Shalamanov holds the title of Associate Professor on automated information processing systems and has more than 150 publications in areas of CIS architecture and development, information warfare, decision-making support, national and regional security policy, military art, defence re-engineering and planning. Dr. Shalamanov is a member and co-founder of the AFCEA Chapter Sofia Chapter. He serves on the International Advisory Board of the Geneva Centre for Democratic Control of Armed Forces. *E-mail*: Shalamanov@GCMarshall.bg

IRENA MLADENOVA is Executive Director of the Economic Policy Institute. She was in charge of organizing the EPI international conference on "Restructuring of Defence Industrial and Technological Base in New and Prospective NATO Members" (Sofia, June 2000), the international conference on "Interoperability of Bulgarian Military Education and Training System with NATO Standards" (Sofia, May 2001, co-organized by EPI and the Rakovski Military Academy), and the EAPC Seminar on "Privatisation and Restructuring of Defence Industries, with a Particular Focus on the Progress Achieved in South Eastern Europe" (Sofia, March 2002, co-organized by the NATO Economics Directorate and the Economic Policy Institute). She worked with an EAPC Fellowship on "NATO Enlargement and the Costs for Bulgaria to Join NATO" (awarded to EPI, 1999-2001). In 2001, within the State Department International Visitors Program she undertook research in the USA on the challenges to Bulgaria in view of its membership in NATO. *E-mail:* epi@bulnet.bg

The New World of Defence-Dependent SME's

Thomas F. SHEA
U.S. Department of Defense, Washington, D.C., U.S.A.

Abstract.
1. The modus operandi in the past for SME's in defence-related work, in the USA and the West.
2. Defence downsizing beginning in the late '80's in the USA.
3. What the Department of Defence did to help affected SME's; Needs.
4. Some conversion examples that emerged from the Starwars experience.
5. What does the future of defence work and defence contracting look like now?
6. The next economy is an innovation-led economy.
7. What is happening in defence spending in the USA today?
8. How do such changes affect SME's in the West, and in South-Eastern Europe?
9. What do SME's need to do to adapt and survive in such a rapidly changing world?
10. Hi-tech industries; low-tech industries. No such thing.
11. Recognition of the rate-of-change in our lives today.

Introduction

For small businesses in the U.S., life as a subcontractor in the defence sector was pretty good. You could be very small, but it was important to remain focused. Your task was to produce a particular product, usually manufactured and often somewhat technical, and supply it to the next level, producer/business, as needed. And when the Federal Government needed, say, a new weapons system, or missile, or rifle, or tank or whatever, when the order came down from the prime contractor, or subcontractor, you were there with your product, your cost, and your production capability. The biggest phrase in your business life was military specification or "mil spec." There was a repetition and orderliness in the process that raised both your comfort level and bank balance.

How cruel, then, when He Who Giveth became He Who Taketh Away, and that is exactly what began to happen in the mid-1980's in the U.S. defence sector spending, which climbed every year as far as the 1950's, began to slow down and billions of dollars were no longer forthcoming, and job losses grew to thousands and tens of thousands, causing unemployment in all sectors of the work force, from hi-tech scientists and engineers, to the lowliest blue collar factory worker. In short, their world fell down around them . . . and most businesses did not have a clue what to do.

For reasons that are probably rooted in equal parts of history and tradition, the United States Federal Government believes it is incumbent to help citizens who have been hurt financially by actions it took, so the economic dislocation caused by defence downsizing, and also by closing military bases, which began in earnest in 1989, triggered Federal Government's assistance. It came in part in a very specific way: through strategic planning.

It was always the position of the Department of Defence not to rush to the scene with a basket of financial assistance; it was more important for the community to step back and undertake two actions: 1) carefully measure the extent and depth of business loss and job loss in the region, in detail; and 2) devise a strategic plan, based on the communities' assets and liabilities, that will lead to new business formation, re-focusing existing business to civilian markets, and create jobs for local residents. Military personnel were looked after by the military; i.e., they were sent to other military installations and kept on active status. This strategic planning exercise, often a first-time event, allowed the community to get a panoramic view of itself as a whole, and address the litany of problems in order of severity. The planning component was executed by an outside professional planning organization, paid for by the Department of Defence. Local leaders and residents were afforded the opportunity to work along with the planning professionals and provide input, and were offered full briefings at the end. Indeed, it was the local community that would accept or reject the new course charted by the planners. Reaction was usually predictable. Local residents and especially businesses welcomed the refreshing approach to a new start, and making a turn in the road.

Meanwhile, SME's were still hurting from diminishing defence work, and dealing with some or all of the following problems: a) no product; b) precious little product-development capability; c) no marketing experience or skills; d) little knowledge of the civilian market, or how they could fit into it; e) how to obtain financing; and, f) where to turn next.

Of course some firms went bankrupt; others simply dissolved and their workers scattered elsewhere, to new jobs, new locations, or retirement. Yet others, most of them, were determined to hold their ground, take a fresh look at new civilian markets, and reach out to local service providers for assistance in two key areas: 1) business management assistance, and 2) financial assistance. Of the hundreds of thousands of SME's who took these steps, there is no one picture to describe what they did, what worked for them, where that trail led them, and how their story ended, happily or not. But examples of how two businesses coped are worth looking at.

- Durodyne Corporation, located in Arizona produced industrial hoses for military and aerospace customers. As late as the early 1990's, Durodyne sold 90 percent of its produce to defence. After working though many of the steps outlined in this paper, company sales to civilian markets reached 60 percent, and growing. It is getting ready to introduce a new line of hoses that will boost commercial sales. Instead of manufacturing quiet valves for submarines, it is producing bearing for commercial helicopters. Investment in a web site promises greater commercial exposure.
- American Tool Service specializes in water jet cutting, laser etching and vacuum forming. Also located in the State of Arizona, 70 percent of sales were in the defence sector, and a temporary workers' agency provided the employees. After seeing no more future in defence work, the company turned to civilian markets, produced specialized cable for use in microwave ovens. Civilian sales have soared to 80 percent.

What we have seen thus far is what happened in a downsizing period that ended roughly in the early to mid-1990's. While that is of historical interest, the more relevant question is what SME's, especially in the defence sector, will need to understand in order to compete and succeed. And related, what effect will the rapid rate of change in our society have on SME's in general, in and out of the defence sector.

1. SME's and the world of defence spending

First a look at SME's and the world of defence spending. The end of the Cold War does not mean a sustained reduction in defence spending, at least in the USA. To the contrary, defence spending, and the budget of the Department of Defence, is actually soaring, $328 billion for the year beginning in October 2002 compared with $279 billion in1999. This increase stems from two major sources -- the war against terrorism; and the creation of a new generation of weapons systems for the U.S. military. Let us look at that.

As reported in the Washington Post newspaper recently, just 10 years ago, a major U.S. defence contractor, Northrop Grumman Corporation was primarily a manufacturer of aircraft; for example, the B-2 Stealth Bomber; also, mechanical parts for fighter jets and Boeing passenger planes. Today, Northrop Grumman has abandoned those lines of production in favour of electronics systems and information technology. Even the language of defence contracting is changing. Not long ago, a prime contractor was one that built a jet or a ship. Now, those vehicles are simply "platforms" for sensors and information systems, according to the Washington Post. Said a former defence official: "the prime contractor of the future is a firm that can integrate onto a platform all the defence electronic and facility connectivity between that platform and others in a system-of-system world." In other words, there will be a military in which all the "nodes" of combat --- soldier, pilot, sailor, general --- are linked in an Internet-style network that lets them exchange information. Add satellites and unmanned spy planes, and soon the average soldier has a godlike view of the battlefield for miles around.

This mega-shift to hi-technology warfare systems will not be achieved by just a handful of prime contractors, but rather through the same historical interdependent network of subcontractors and suppliers of the past, but with new products and technology. The opportunity for SME's is enormous, and the ones that can adapt to this new order, whatever part of the world they are in, can reap great benefits. It has often been said that there is no such thing as hi-tech or low-tech industries; only hi-tech or low-tech businesses. It is their choice. SME's have often been characterized as flexible and quickly adaptable to change, sometimes compared to gazelles, those small, graceful and swift Asian antelopes. But as in history, there will be dinosaurs too, those SME's that cannot or will not step into the new economy, an innovation-led economy. Let us take a look at that new economy, and how it will affect SME's everywhere, including South-Eastern Europe. To set the context of change, consider the following statement: "When the rate of change on the outside (of your business) exceeds the rate of change on the inside, the end is in sight." Is the end in sight for the SME's we are discussing? If they resist change, it is in sight. Consider this:

- Resource-based economy: the past
- Technology-driven economy: the 1980's
- Innovation-led economy: the future.

If SME's wishing to continue pursuing defence-related work, the foregoing presents a picture, admittedly long-range, of what the defence weapons mosaic will resemble. Their challenge is to adapt their business capabilities to meet those changes and fill that role. But most SME's, bouncing off a flat defence-oriented market, are choosing, even being forced, to find a new direction, to civilian markets, and with new products and services. We need to look at what the characteristics of that new marketplace will be. What are the growth areas of the future? According to a person who studies these things, Edward Barlow, they are:

- Defence
- Security

- Bio-technology
- Healthcare
- Law enforcement and corrections
- Hi-tech manufacturing
- Education
- Government
- Construction
- Insurance
- IT

While these categories are primarily for the USA, they will apply selectively to other countries worldwide, and SME's need to study them carefully to figure out how they can become productive in any of them, and measure that against local and regional (even national) interest. That is the first step in pondering out one of life's hardest questions: in what direction do I go?

2. The next step

Let us proceed to the next step an SME will encounter. Let us further assume it has selected a category from above, hi-tech, not manufacturing necessarily, just hi-tech in general. What are the ingredients for high tech growth? Again, according to Mr. Barlow, they are:

- Access to venture capital
- University research & development programmes and heavy R&D spending
- Abundance of highly skilled workers
- Clustering of similar high-tech companies
- Access to reliable power
- Quality of life.

While again, these elements were ascertained for high tech in the USA, it is helpful for SME's to step back and take another look at basics, some of which have been referred to in earlier paragraphs. It is axiomatic that SME's, or any size business for that matter, need a well thought-out, understandable business plan. There is more business plan formats than one can count, but they all get down to fundamentals: what business are we in? Who are our competitors? Who are our customers? How do we market our product (service)? What are our working capital and financial requirements? From where? Cash flow projections over specified time periods? etc., etc., etc. SME's, especially those turning to civilian markets from defence, need to undertake careful study and analysis of the market today, and in the short- and mid-term future as best possible. The all-important notion of marketing, actually writing a marketing plan, is too often overlooked or taken for granted, and SME's realize too late that they are being left behind by the competition, and are too weak to catch up. The same can be said for understanding SME's financing requirements, now and in the future. The more correct phrase is early-stage financing, that is, funds to take the product or service idea into some form of reality, enough to be market-tested. While such financing may be better called "loans," it is speculative and risky, but it is that early-stage money that is indispensable for the civilian market-oriented SME. Indeed it would be accurate to compare the condition of the SME's in question to a start-up small business, for indeed both need the components alluded to earlier: new product, product-development capability, skilled marketing ability, market insight and knowledge, their market niche, and financing. Underpinning all of this is the commitment from so-called top management that they are serious and determined in this new direction.

3. Conclusion

The picture that emerges, therefore, is that SME's forced out of the defence sector and turning to local civilian markets, are not particularly unique, nor are their problems. These SME's are facing situations which, reduced to fundamentals, have been faced by countless SME's before, either ordinary start-up SME's, or SME's wherever, that figured-out the system and how to cope and survive in it. The good news is that some of them prospered and that formula is fully relevant to SME's located in South-Eastern Europe.

To end this paper, presented herewith, again from Mr. Barlow, are examples of how particular items have changed from the Old Economy to the New Economy.

Old vs. New Economy

Item	Old Economy	New Economy
Compete	Physical assets	Intellectual assets
Profit	Control cost	Add values products and Services.
Work force	Job-specific Skills	Transferable skills; Lifelong learning
Geography	Near resources	Near collaborators and competitors to boost innovation
Competition	National	Near collaborators and competitors to boost innovation
Capital	Debt Financing	Venture Capital

The "Economic Missionaries" – a possible solution for global integration of SME's from Central and Eastern Europe

Mircea SLANINA, Ph D
Fundatia CENTIS,
Iasi, ROMANIA

Abstract. Even though a country has demonstrated the democratic character of its political system it is not enough to fit the requirements for NATO membership. Good economic dynamics are a fundamental feature and such a requirement became a component of NATO policy versus that of candidate countries. But the Central and Eastern European (CEE) economies are still different from the economies of other member states. The reconstruction of CEE economies could be a testing field for the NATO philosophy versus recent entrepreneurial trends.

A decisive evolution of the contemporary world occurred in the entrepreneurial environment: it is the small and medium enterprises (SME) phenomenon. As a living organization, NATO has to naturally react to this challenge. Given its importance, NATO's option could only be a long-term choice. The citizens of the CEE countries would be very impressed to see NATO itself becoming involved in the restructuring of their economies, trying to cope with the delicate issues of the SME sector.

The author provides some examples from his own country, Romania and suggests a possible bottom-up approach for the integration of CEE economies in the global market.

1. A challenge from the entrepreneurial environment

In a simplified view, the entrepreneurial world today is strongly polarized. The **globalization** does represent the most visible trend. The large multinational corporations are its promoters and have become the most influential actors in this movement. In the same picture, the opposite trend is represented by the growing importance of SME's, by their natural creativity and by their specific attention to the local/regional needs of communities. An excessive polarization of any economy in a few strong corporations and many powerless companies could become dangerous.

It is a common and a normal decision: the main NATO providers of technologically intensive products or of very complex equipment are the large corporations. Is NATO a prisoner of those corporations? The generally accepted answer is no: SME's from the member countries are frequently providers for NATO, including some high-tech products or services. In the apparent dilemma 'globalization versus local communities', NATO - as a democracy promoter -, should never neglect local, regional, or national aspects.

As a generally accepted metaphor, the SME represents the human-size tailored economy. That's why the long-term persistence of the SME phenomenon in all the economies is free of any doubt and NATO's interest in SME evolution in the candidate countries is unavoidable. But for now the features and policies of SME's are completely different in member states and in candidate countries.

2. The SME as NATO provider in a Western country

Of course, often the SME cannot compete and deliver most complex systems or products. However, the SME networks that form the Western economies are able to deliver important parts for complex systems, for military or civilian markets. In such networks, almost every SME is highly specialized. Sometimes they deliver directly, but, usually, their customers are the prime contractors. Commonly, at the same time, the SME is part of many networks or clusters or at least supply chains - be they formal or informal. The SME's compete less directly, more as members of networks and often the hard competition develops between the alternative networks, consortiums, clusters, associations, etc.

The network became more important than the company and the quality of interaction more important than the size. In such an economic fabric the prime contractors preserve their specific and essential role. They issue the orders, finally they assemble the work and they preserve the specific discipline in the 'systems of systems'. The difference in size grew into a welcome complementarity. In their own way, all the actors in the economic activity influence and contribute to the local/regional/national development.

Every SME preserves a large autonomy; meanwhile, the business environment protects it as a necessary element of the regional economic fabric. The SME management keeps the balance more easily between the customer's and the network's requirements and naturally considers the SME's social role. The remarkable ability of SME's to be adaptive and innovative at the lowest costs are highly valued by the social and economic development participants. Competition was replaced by 'coopetition'. Gradually, in a conscious or unconscious manner, the networking (or, better said, the ***bio-thinking***) became the dominating business philosophy.

The scheme presented above is close to an ideal one; even so, the SME's situation in the ex-socialist countries is very different. A successful integration of the candidate countries' economies into international economic cooperation would be possible only through a good understanding of the local manager's and entrepreneurs' mentality and habits. Let us start the analysis with an unfortunate example of 'strange' networks.

3. A real story: "the colonels' economy"

In some of its features, Romanian socialism was unique. It is worth remembering Ceausescu's absurd ambition to pay all the country's debts in a few years. Romania tried to design and manufacture all the products it needed, including military equipment. As a result of this decision, the country was surrounded by something similar to the Chinese Wall.

In a centralized country, the R&D institutes owned a large part of the national scientific and technological knowledge. The "no more imports" decision meant frequent contacts of R&D engineers with military officers of significant rank. Diverse products and important equipment started to be manufactured in small series, many times with the contribution of institutes.

After the Romanian revolution a decisive change in our people's mentality occurred: many persons started to dream of how to hit a strike and get rich. As a result, numerous and strange business alliances developed. Strange (informal) associations developed between the new SME's - set up by some R&D engineers - and certain army colonels. The colonels provided important military procurement orders. Bribery was the name of the game.

Such "joint ventures" distorted and still distort the market because the traditional military staff covers all the army-related information with a shield of secrecy. For the corrupted personnel it is the perfect protection for their selfish interests. On the contrary, the majority of SME's would prefer a fair competition in obtaining civilian or military orders. This type of government bureaucrat does not like transparency, he is not interested in innovation or in technology transfer. The bureaucrats will never stimulate alternative networks as military material suppliers. Now, because of defence budget downsizing, such technologically advanced manufacturers are barely surviving on the civilian market. Some of them could be revitalized in a better business environment.

The bottom line of such a story is: the structure and practices of ex-socialist countries' economies are very different to the business style in Western countries.

4. NATO and the candidate countries' economic environment

Despite corruption-affected economic relations, in the Romanian business environment today, competition has a primary role. But the economy is still non-transparent, unstructured, without representative organizations for every industry. The numerous newly-founded SME's have not yet developed into large, viable companies. The majority of the State-owned companies have either split or are in poor condition. In fact, the majority of actors in the economy are SME's. An ex-socialist economy like Romania's is actually a chaotic conglomerate of non-focused SME's. That is why in contacts with economy actors of CEE countries the specific psychological features are peculiar.

To define the State's role in the economy, it is important to first consider the public administration image in the business environment. Generally, the small entrepreneurs are not confident in the professionalism, the good will and the honesty of the public administration at all levels, regardless of the political parties that support the government. The people no longer trust the helpless and corrupted new regional development agencies, which, in fact, do not fight for decentralization. (However, these regional administrations could play a positive role in building a regional specific in the national economy.) Consequently, for the integration of CEE economies in large international organizations like the EU or NATO, the top-down approach is not the best solution, and not even necessary. Direct economic contacts are much more recommendable.

Before such contacts, Western visitors have to be warned about the underdeveloped business culture that has proved to be the most resistant feature against positive change. During the last two centuries, the centralistic thinking was inoculated in all Romanians. So, the main enemy of the private initiative is inside people's heads: this is the unconscious fear or belief that any initiative will be an unsuccessful one, if it is not started by an order that came down the hierarchical chain. This is the paradox of the ex-communist countries: after 50 years of forced collectivism, there is a serious lack of an associative spirit. In many respects, we can appreciate that Romania follows some models of 19th century capitalistic development. The "Hollywood model" of the entrepreneur – strong, lonely and an enemy of any public authority – is now the most common model. After so many economic experiments all over the world, the Romanian way could be assessed as an attempt to build here a ***primitive capitalism.***

Such a negative trend is not a fatality and NATO's enormous prestige should be pro-actively used for CEE countries' economy restructuring. Privatization is mandatory and urgent, but the privatization itself will not automatically generate a modern economic fabric. Even the privatization is not yet exhaustive. We have to carefully look now at local SME integration in the global economy. But who would be able to do that and how?

The right way to restructure such economies is to build networks, well-oriented to the needs of the main customers. It means the stimulation of an aggregation of Eastern SME's into strong clusters, able to provide competitive products and strongly connected with the global market.

The Western SME's do not have the financial power and are not strongly motivated to act remotely in achieving such a goal. This is more evident when they are interested in new, innovative developments; to guide a remote innovation process is too complex a task for them. Only the prime contractors – no matter whether for civilian or for military equipment – could play a significant role in this restructuring. In principle, they are more interested and able to integrate new providers within their networks, which use qualified and cheap labour. So, a right policy for NATO's prime contractors would be very important if it would start as follows:

- to encourage the specialized/regional networks,
- to promote transparency in their business relations with SME's,
- to stimulate associative behaviour (transnational included),
- to foster the bottom-up building of networks,
- to act according to sustainable development rules.

But even the large Western corporations do not possess all the abilities to cope in an efficient way with the specific features of Eastern economies. It is not enough to identify the proper SME's because of the dominant, very individualistic, philosophy of the local "business heroes". Additionally, in building the networks for them it is worthless to interact with many isolated SME's; a local professional would perform better in his country than a foreigner.

5. "Economic missionaries" – an auxiliary but important means

Often we can find suggestions in the past for present problems; a creative adjustment to conditions today could be the key for their new success.

Some hundreds of years ago, the geographical revolution, followed by the commercial revolution, anticipated globalization. At that time, the nations promoting globalization sent missionaries to their future colonies. The missionaries "exported" culture (mainly religion), made useful contacts and gathered information. They accomplished their mission at very low costs.

We think Western corporations have now to consider a type of "economic missionaries" to extend their networks in the emerging economies. The missionaries have to be recruited, prepared and later used as "crystallization seeds" or at least as liaison officers for multinationals, including the NATO prime contractors. However, we strongly recommend a new approach for such a policy: the missionaries have to be selected from among the professionals of Eastern countries. It is mandatory to choose them from the same industry as the industry of their employer, from among the very well-known professionals of that field. Every "missionary" has to spend a working stage within the middle management of the employer. He/she has to be involved in different key positions of the networks, clusters or supply chains, to understand and to learn networking. The stage duration could be set for several months or even years, depending on the industry, strategies and individual characteristics.

According to the contract, after the stage, the "missionary" should act in his/her country for network building, inside the similar industry of the country. (Of course, the "deliverables" of such a contract are now hard to define exactly.) Gradually, the newly-built supply chain has to be connected to the prime contractor network. To help their "missionaries" in developing the networks and in promoting the NATO economic policy versus that of candidate countries, the prime contractors should include in the tenders' files some clauses favouring specialized clusters, regional networks or innovative SME's.

It is likely that similar procedures did work in certain countries and industries, but accidentally, not as a systematic policy. It is probable that such a procedure would contribute to a harmonious integration of emerging economies in the continental and global business environment.

Offsets as an Alternative Tool to Support Innovation and Institutional Development in the SME's

João Pedro Taborda,
Inteli, Portugal[i]

Abstract. The paper presents a framework to analyze the Efficiency of Offset Programs when these are used as an alternative tool to finance innovation, institutional development and change in the local industry. Particular attention is given to the involvement of Small and Medium-sized Enterprises (SME's) in these offsets activities.

Traditionally, Offsets are defined as the commercial contracts between the prime contractor[ii] (the *offset obligor*) and the government doing the acquisition, in order to compensate the local economy against the significant amount of public resources spent in the acquisition of sophisticated equipment for military purposes.

Moving from a stage where these agreements have been focused on commercial assistance in exporting existing products and services, most countries are inviting nowadays their prime contractors as *offset obligors*, to submit activities in the compensation projects[iii] with a strong innovative content, and maximum involvement of the local industry. This can be done not only by creating partnerships with local big companies, but also through activities involving SME's directly, as partners in the offsets projects, or indirectly as suppliers to these local partners.

Beyond these requirements, and in order to preserve a fair return that should be associated with the spending of significant resources coming from the tax payers, *Additionality* should also be addressed. Nations demanding offsets express a concern that the new activities where their industry becomes involved are a direct result of the Offsets Program, and do not represent transactions that would take place in the regular course of business (Udis, 1994).

Introduction

By using offsets as an alternative tool to support *Innovation*[iv], governments managing Offsets Programs[v] look now and to the future to interact with the offset *obligor*, stimulating the creation of new partnerships between local companies, including SME's (as *offset recipients*) and foreign entities (as *offset sources*[vi]). Priority is given to activities that stimulate the absorption of new competencies by local supply chains, and its further integration in innovative solutions, both at the *product/services* and *process* levels. Technology transfer is one of these priorities, with nations welcoming long-term business and partnerships supported by the consolidation of interaction routines between the *offset recipient* (local company) and the *offset source*.

Information provided by the United States Department of Commerce on the approach followed by 25 countries shows that, among these, fourteen are explicit in establishing *Technology Transfer* as one of the main priorities for eligible offsets activities. Data from the same source shows that, between 1993 and 1998, 11% of the amount of total transactions was generated from this category, behind *Purchase* (37%) and *Subcontract* (29%).

This paper presents the results of a research where 17 international cases of Offsets Programs were analyzed. In order to understand some of the factors influencing the use of offsets as a tool to stimulate innovation, technology transfer and institutional development, a model is presented, providing a framework to evaluate the *output* of each case (*Case Efficiency*) of Offsets Programs. For research purposes, and based on the existing theories of institutional development and change (North, 1994), *Efficiency* was defined as the level achieved in each case in the implementation of activities involving local SME's, which contributed to the creation of new institutions favourable to the absorption of competencies producing a positive impact on their competitiveness in international markets. Higher efficiency rates were given to cases where the Offset Program was able to provide local companies and industrial SME's with new competencies that made possible their integration in global supply chains. Special attention was paid also to activities related with SME's operating in sectors of higher technology intensity[vii], as providers of high value-added products and services.

As the *input* for the model, qualitative and quantitative analyses were performed on the impact over the *Efficiency*, of a set of variables including: i) Volume/Frequency of the acquisition; ii) Payment Associated (existence of penalties); iii) Kind of local investment; iv) Type of activities (direct foreign investment, technology transfer and export assistance); v) Kind of agreements; vi) Level of local competencies; vii) Social support towards the acquisition; vii) Visibility of the process, and viii) Project focus defined by the purchasing country (*activities-players* or *players-activities*).

The conclusions resulting from the case analysis show that structural benefits are generated from Offsets Programs when local companies learn to do *differently* and *better* (a *Specialization* effect), and activities performed under the offsets projects take place beyond the borders of the local companies directly involved (a *Difusion* effect). Based on the analysis and evidence provided by the cases, this text supports the importance of combining a *Specialization* effect with a *Difusion* effect, as a key aspect to generate multiplying benefits for the local economy.

1. Aspects on Innovation and Technology Transfer in SME's

Beyond the conventional definition as a process of identifying new products and services with market acceptance, innovation is also seen as an attitude to welcome risk, based on a mindset and way of thinking focused in the present and future (Kuczmarski, 1996).

Technology plays an important role in the development of this mindset in order to make possible the implementation of new products and services that may provide new competitive advantages. In the way to achieve this goal, other factors must also be taken into account such as access to capital, access to markets and management of proprietary rights (David, 2000).

The diffusion and global use of communications technologies has increased the speed and has reduced the cost in access to new information, providing also the opportunity to develop new solutions faster (Confraria, 2001). Levels of specialization increased in industries with a higher degree of technological intensity such as automotive, aerospace and electronics, where *modularization* became a trend, inducing new types of business relationships between the Original Equipment Manufacturers (acting more as integrators) and their suppliers. From a *build-to-print* scenario, suppliers moved to a stage where their clients are demanding new competencies in areas such as product development, quality and logistics, becoming more dependent on information (Albrecht, 2000).

As a result, specialization increases through the continuous access to new information, with the same product being manufactured using an increasing number of technologies, and the same technology also being used in the production of an increasing number of products (von Tunzelmann,1999).

In managing innovation effectively under this business environment, technology transfer must be analyzed in terms of the balance between the three forms of technology: material, non-material and documented, and according to its source (Pavitt, 1984). For an organization to engage in innovative activities able to bring new competitive advantages, the balance between the three forms should be considered. With a growing demand in specialization, the access to material technology (equipment) and documented technology (proprietary rights) may not be sufficient (Nelson, 1997), suggesting the importance around the creation of routines that may facilitate the exchange of tacit knowledge (Nonaka, 1991).

For the SME's, this dynamic presents them with the need to implement these routines as a key element in the process of sustaining a strong competitive position in the market. The consolidation of local supply chains, with the interaction among the different entities providing a framework for the implementation of these routines, has been a priority in countries like the United Kingdom[viii], helping the SME's to face current challenges such as:

- Business environment with increasing competition (Globalization)
- Search for economies of scale
- Resource Optimization
- Search for new opportunities in new markets
- Product and process innovation under a common framework for the regulation of competition

These facts raise the need among SME's to invest in information and technology that may leverage their knowledge, integrating it into new ideas with wide acceptance either in existing or in new markets. From the policy perspective, the local government should coordinate these activities in order to assure economic growth. Experiences by countries dealing with offsets show that SME strategies and innovation polices need to find the right balance in this coordination. Considering the example of Portugal, some authors present as a dominant strategy to consider the ability to create, manufacture and distribute final products in international markets, as a way to focus specialization and interaction, not only among the companies, but also between them and the scientific and technological system (Neto, 2002).

Beyond the process of technology transfer that may support these kinds of priorities, other factors, which are not analyzed in detail under the research, may also play a relevant role when offsets are used to stimulate economic growth (Figure 1.). That includes the educational, financial and judicial systems (Neves and Rebelo, 2001).

Technology transfer may increase with the level of Technology Intensity, which is higher for sectors investing more in R&D activities. According to Kinder and Lancaster (2001), the ability of the SME in receiving and integrating new technologies into products and processes is related to the establishment of routines that enable the selection, analysis and validation of external information. The same authors call it *Absorptive Capacity.*

The *Absorptive Capacity* should be considered as an asset that enables an SME to capture technology and knowledge in their three forms. The effort of selecting information from outside the company may have, as a main result, an increase in the level of specialization through the absorption of *new knowledge*. Technology Intensive sectors need more agility in absorbing this knowledge from outside the company (both in its tacit and explicit form), and integrating it into new business ideas with market acceptance.

This search for *new knowledge* by the specialists in SME's under the scope of offsets projects turns the selection and integration of new ideas into a process, based on a complex system of human interaction among different parties involved in the project: the *offset recipient*, the *offset source*, government representatives and consultants.

Based on the research carried out in seventeen cases, this paper tries to address some of the factors that may be related with the use of Offsets Programs as a mechanism to support technological innovation in the SME's, where this same support may take place directly or indirectly. In the first case, the SME may become a partner in an offsets project, interacting directly with the offset source. For the second, the SME may be part of a local supply chain working for the local partner (the *offsets recipient*), that can also be another SME or a big company.

2. Theory of Institutions, Institutional Change and Offsets

Under the author's approach for the theory of the impact of institutions and institutional change in economic performance, North (1991) provides a framework that was used in this study in the understanding of countertrade and offsets as an alternative tool to foster innovation among the SME's.

Institutions are defined as the humanly-devised constraints that shape human interaction. Institutions structure incentives in human exchange, whether political, social or economic. In simple terms, institutions are the *rules of the game*, providing the structure for everyday life. Organizations (where SME's may be seen as an example) are part of the institutional theory. They are groups of individuals bound by some common purpose to achieve objectives.

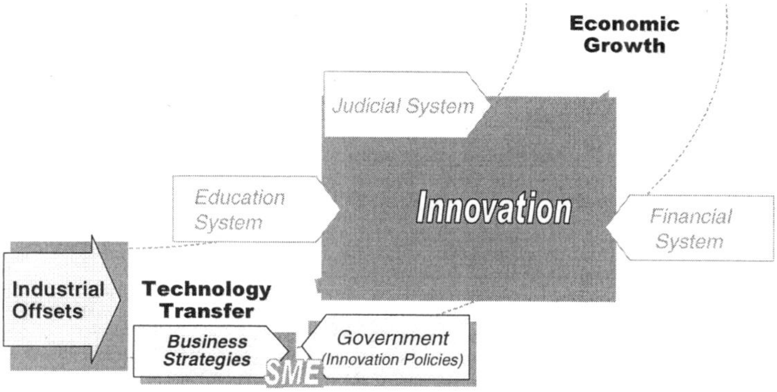

Figure 1. From Technological Innovation to Economic Growth

The major role of institutions in a society is to reduce uncertainty by setting a framework for human interaction, which may be more or less efficient in different areas such as economic, political and social. Institutions have two different forms: formal and informal. When compared with a world of no institutions, institutions try to reduce the costs in human interaction under conditions of limited information.

Institutions interact with organizations and people. The former influences the latter and *vice versa*. According to the author, formal constraints include political (and judicial) rules, economic rules and contracts. Although very important, formal rules play a smaller part in the set of constraints that shape human choices. Informal constraints are culturally-driven. They are the codes of conduct, norms of behaviour and conventions, coming from socially transmitted information. Due to their particularities, informal constraints play the more important role and are more demanding as far as the time to change them is concerned.

Changing formal constraints takes less time than changing informal constraints, which are transmitted across different generations under this context. Culture is defined as the transmission from one generation to the next, via teaching and imitation, of knowledge, values, and other factors that influence behaviour.

Institutional change comes mainly from changing the informal constraints, making the process of change incremental. As a result, learning takes place according to the direction that maximizes individual gains.

There are five propositions to consider in institutional change:

1. Organizations interact under an environment where offer is above demand, raising competition as a key issue;
2. The existence of competition forces organizations to invest in knowledge and learning on a continuous basis;
3. The institutional framework defines the kind of knowledge that may bring more revenue for the organization;
4. Perceptions result from the mindset and attitude of the players;
5. Issues like economies of scale, complementarities and existing networks make institutional change an incremental process.

In studying how offsets can contribute to institutional change and positive economic performance, these conceptual issues are taken into consideration in the specific environment of Offsets Programs involving SME's. Taking the theory of institutional change as the framework for the research, four aspects related to the dynamics of offsets should be considered in order to provide the definition of *Efficiency of an Offsets Program*, to be used in the case analysis:

- Offsets Programs are usually associated with purchase of high-technology goods, the development and manufacturing of which are based on the integration of different technologies.
- The purchase of these goods provides the opportunity of interacting with global players from sectors with a higher degree of technology intensity.
- The offsets contract has a long-term range, usually associated with the payment of the purchased goods, enabling interactions between offset recipients and offsets sources to take place over a long period. This may motivate the creation and consolidation of norms of behaviour based on an attitude of 'learning with the best'.
- Definition and implementation of offsets projects involve complex systems with intensive human interaction, including official supervisors, managers, technicians and consultants.

As a system highly dependent on human interaction, formal and informal constraints play an important role. Examples of the former are the regulations for the bidding process, and the content of the offsets contract. Informal may include, for example, the attitude towards the Offsets Program itself. From the local companies we may find behaviour based on the assumption that 'We should not waste time with offsets. Leave it to those with the *good contacts*', or 'We are in the package; the seller is the one with the problem, he should tell us what he wants us to do'. From the seller's side an example may be 'Let us take the shortest legal way. Forget the long-term, we have to solve the offsets now and fast, and move to the next one'.

Under these considerations, the priority of the research was to identify the factors producing an impact on the way an Offsets Programs should be managed in order to foster innovation, technology transfer and institutional development. Activities under offsets activities may take place, in some cases, for around a decade, providing the timeframe to change informal constraints and to build new learning routines where individual gains can match collective goals. Chapter 3. provides the analysis performed under this framework in seventeen cases.

3. Analysis of International cases in Offsets Programs

A model was developed for the analysis of seventeen international cases of Offsets Programs. For the present research a *case* is an Offsets Program, although in future studies the goal may be to apply an evolution of this tool at the project level. An *Offsets Program* is defined as the group of Offsets projects considered under the purchase of certain equipment, and implemented under the responsibility of the offsets obligor. Usually, this role is played exclusively by the supplier of the equipment, although some exceptions may occur. This is supported by the case involving the purchase of F-5 from Northrop by the Swiss government. Northrop was the prime contractor (equipment suppliers), but the responsibility in fulfilling the offsets contract was shared with General Electric (the engine supplier) and the Department of Defense of the United States Government.

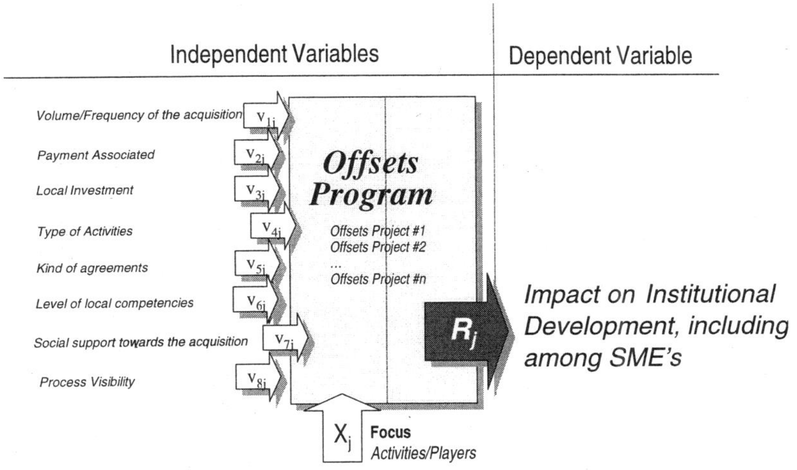

Figure 2. Tool used in the analysis of cases

The system for the analysis includes nine independent variables, and a dependent variable measuring the efficiency of each case (Figure 2.). Under this approach, a definition for *Efficiency of an Offsets Program* is presented based on the framework provided by the analysis of institutions and institutional change described in 2.

For the purpose of the research, the *Efficiency of an Offsets Program* is defined as the existence of evidence showing that some projects implemented under the Offsets Program provide a contribution to an institutional change that supports innovation and economic growth. This could take place either through the consolidation of existing constraints (formal or informal), or through the creation and development of new institutions based on new business relationships. These, taking place under the scope of the offsets projects, may provide external information to be absorbed and integrated in innovative products and services by the local companies, including SME's.

This option was based on the perspective by Udis (2001), who supports the analysis of the efficiency of an Offsets Program taking place on a case-by-case approach, meaning that the same Program may host, at the same time, *successful* and *unsuccessful* projects. In order to simplify the approach followed for the case analysis, a *good case* is a Program where the data provided by the sources shows evidence of projects generating long term effects on the local industry[ix].

A methodology for gathering data and information for each international case was adopted based on a systematic use of primary and secondary sources. Among the seventeen cases analyzed, seven involved Portuguese companies (as offsets creditors in six of the cases[x] and offsets obligor in one of the cases[xi]), where access to primary sources was used as the main priority, since published information was limited to a working paper and news in the press. Unpublished information was gathered through personal interviews with top managers and technical staff working for the companies, as well as reporters, representatives from the Portuguese Offsets Committee (*Comissão Permanente de Contrapartidas*) and consultants who took part in the process.

The analysis of the 10 non-Portuguese cases was based mainly on data and information gathered from literature. Better precision may be achieved in further research work by accessing an increasing number of primary sources related with the Offsets Programs. This will enable new validations from the model, based on the use of more data and information supporting each of the independent and dependent variables.

3.1. The Dependent Variable

The dependent variable, R, is related with the *Efficiency of the Program*, measuring the impact on institutional development of the offsets activities. The value for each case was defined according to the evidence around the existence of projects with a structural impact on the local industry, generating benefits beyond the borders of the organizations directly involved in offsets projects from the Program, and beyond its timeframe.

For each case, the research methodology was oriented in order to gather data and information supporting the existence of benefits resulting from the projects such as:

- New products and new manufacturing competencies, based on partnerships with the offset obligor or its subcontractors;
- Launching and consolidation of local organizations with new and distinctive competencies;
- New business relationships, extended in time beyond the conclusion of the Offsets Program;
- Development of local supply chains through the involvement of local companies beyond those playing the role of main contractors in the offsets projects;
- Access to new markets by local companies or the overcoming of existing barriers;
- Changes in attitudes or assumptions related with other cultures or behaviour regarding the use of offset projects to develop new activities with strong innovative content.

This information was used to assign a value to the dependent variable, R, according to a scale, rating each case from (-1) to 3 (see Table 1.). The scale considers two different effects associated to each project: the *Specialization* effect and the *Difusion*[xii] effect. Under the framework used in the current research, the structural benefits for the local industry associated with an Offsets Program are based on the combination of these two effects (Figure 3.).

The *Specialization* effect is associated with knowledge acquisition by the offsets recipients (that may include SME's), and their ability to learn in order to do *differently* and *better*. To achieve this, the local company taking part in the offsets project will implement internally new learning routines, enabling the generation of innovative ideas that may provide new opportunities under the multiplying effect to be generated. A stronger absorptive capacity provides the organization with the capacity of selecting external information under the scope of the offsets project. The need for an effective absorptive capacity increases with the degree of technological intensity of the industry where the Program and projects take place, so the *Specialization* effect may tend to produce stronger benefits when companies from these sectors become involved.

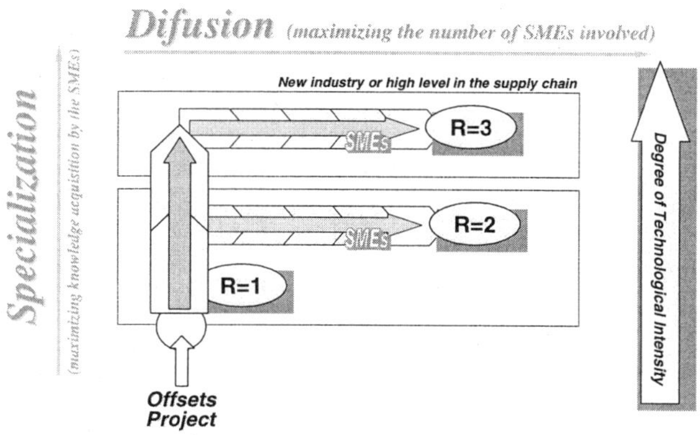

Figure 3. Combined effects in the measurement of case efficiency

As a benefit, this effect may enable a company to move to higher levels in the supply chain (for example, from parts' manufacturer to part or system designer), or even to move up, into sectors with a higher degree of technological intensity. Although very important, the *Specialization* effect should not be seen as an end result when an Offsets Program is managed as a tool to foster technological innovation involving SME's.

Table 1. Scale used for the Dependent Variable, R

R = -1	Negative Impact on the existing institutions;
R = 0	No impact or minimal;
R = 1	Positive impact on institutions towards technological innovation, but involving a small number of players (*Specialization*, but no *Difusion*);
R = 2	Positive impact on institutions, involving a larger number of players, including SME's. *Specialization* with small impact on supply chain position or capability to enter new sectors with higher degree of technological intensity, and *Difusion*;
R = 3	Creation of new institutions involving SME's, in a positive direction towards technological innovation (*Specialization* with impact on supply chain position and/or acquisition of new capabilities to enter new sectors with higher degree of technological intensity, and *Difusion*)

In order to disseminate the benefits beyond the borders of the organizations taking part in the projects, the *Specialization* effect must be complemented by a *Difusion* effect aiming to maximize the number of local SME's involved. An expected benefit is the creation and consolidation of local supply chains in sectors with a higher degree of technology intensity. This will depend on the volume and time range associated with the offsets projects. If the offsets project provides only a discrete opportunity in time, routines among the members in the supply chain, including SME's, will be more difficult to establish and to consolidate.

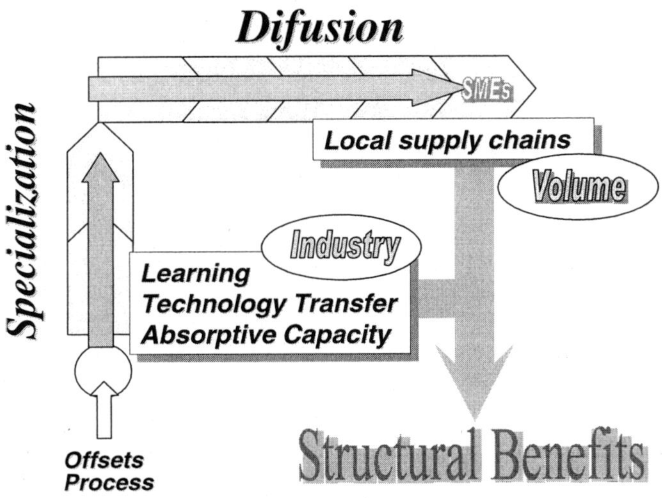

Figure 4. Structural benefits of an Offsets Program

3.2. The Independent Variables

From a qualitative analysis of 33 international cases (later consolidated to the group of seventeen in Table 2.), a group of nine independent variables emerged, the impact of which on the dependent variable was later evaluated under the analysis of 17 of the 33 cases.

A brief description of each of the variables is presented. The associated scales are presented in Annex I.

X_j. Focus. 1: Activity-focused; 0: Player-focused

According to the available information on each case, the variable aims to verify if the prevailing rule was deciding first 'what to do', before 'who is going to implement it'. For cases where X=1 it was possible to gather data and information that could support an orientation for certain types of activities, or for the development of certain industrial capabilities and competencies.

v_1. Volume/Frequency of the acquisition

A value was assigned to the variable according to the volume and attractiveness of the buying country as a client. In absolute terms, a certain purchase may not be significant but the value for this variable may still be high, if its origin is a country under a geo-political situation that provides guarantees of future business for the prime contractor.

v_2. Payment Associated (Penalties)

In some cases, the payment of the purchase to the prime contractor may be associated to the good execution of the offsets contract. Penalties may exist and this is becoming a common practice today, although cases from the past (Case 1, Northrop and Switzerland) took place under *best efforts* agreements, with no penalties involved.

v_3. Kind of local Investment

Beyond technology transfer, purchasing and subcontracting, governments try to use offsets as a tool to attract foreign investment to implement the projects. It may take place through different ways, including direct support to the project's activities (financial or assets like people and equipment), *greenfield* or joint ventures with local companies.

v_4. Type of Activities

Offsets projects may include different types of activities such as purchase of goods and services available locally, subcontracting, intermediation for new businesses, access to licences and other type of proprietary rights (including training) and transfer of new technology.

v_5. Kind of agreements

Offsets may be direct or indirect, according to the final target of the main activity. If the project aims to provide a product or a service to incorporate in the production or operation of the equipment under purchase, the project will fall in the category of direct offsets. In each of the cases, there are financial flows involved. Other two types of projects exist where that is not the case: barter and buy-back agreements. In the former, goods are exchanged directly. Under buy-back agreements, a foreign investor is responsible for the activity locally (human, equipment and financial resources) being paid with the goods under production[xiii].

v_6. Level of local competencies

This variable assigns a value according to the level of development of the local capabilities before the implementation of the Offsets Program. Countries like Turkey and Spain used a Program to start an industry that almost did not exist (Hammond, 1990). Others like the United Kingdom and other NATO members manage the Program in order to compensate and leverage the technological capabilities of local prime contractors and associated supply chains[xiv].

v_7. Social support towards the acquisition

Albeit its recent use in humanitarian missions, the purchase of military equipment has always been associated with the threat of war. According to the threats in the particular region, local support from public opinion to the acquisitions may have an impact on the efficiency of the Program.

v_8. Process Visibility

Evidence from the work carried out under this research shows the existence of very limited information concerning Offsets Programs and their impact on the local industry (Martin, 1996). Offsets projects have big volumes, and for that reason also, are surrounded by secrecy, which, associated with a general lack of knowledge about the tool by the public opinion, usually raises *grey clouds* over the process of managing the Program. The variable assigns a value according to the visibility that was given to the Offsets Program, including activities under development, players and benefits for the local economy.

3.3. Using the tools in the analysis of the 17 cases

Based on the analysis of 17 cases and data from other 16 (where data was considered as insufficient to build a case study), several questions are addressed:

- From the evidence of each case, what are the factors to consider in an Offsets Program used as a tool to support innovation?;
- What should be the priorities in terms of industrial sectors involved in the offsets projects?;

- Which kind of activities may be the most appropriate, based on the definition used for the 'Efficiency of an Offsets Program'?;
- What are the aspects that should be considered in the analysis and evaluation of the players implementing those activities?; Should the players be chosen first, or the activities to be implemented through the projects?

According to the data and information available, a value was assigned to each variable in each case. Table 2 shows the results obtained, as well as the accuracy level for the data, based on the type of sources. Cases with accuracy '3' were analyzed from information gathered from more than one source with scientific value. Accuracy '2' was based on one single source with scientific value and personal interviews, and accuracy '1' based on personal interviews or sources such as business news.

A statistical test, using linear regression methods, was performed on the data generated resulting from the analysis of the seventeen cases (Table 2.). A co-relation between the variable X and the dependent variable, R, was identified, with a statistically significant 91.5% level. This result suggests that the *Efficiency of an Offsets Program* may be higher when the priorities around the activities to be developed under the offsets projects are defined before the assignment of the players; meaning the definition of "what to do" before "who is going to do it".

Table 2. Results from the application of the tool to the 17 casesXV

Seller/Buyer	X	V_1	V_2	V_3	V_4	V_5	V_6	V_7	V_8	R	Accuracy
1. Northrop and Switzerland (1975)	1	2	-	0	0	4	2	1	0	2	3
2. General Dynamics and NATO (1975)	1	3	-	1	3	2	1	2	1	3	2
3. General Dynamics and Israel (1980)	1	3	-	5	3	3	1	2	2	2	3
4. McDonnell Douglas and Japan (1981)	1	3	-	1	3	2	2	2	1	2	2
5. McDonnell Douglas and Canada (1982)	1	2	-	5	3	3	1	1	1	3	2
6. McDonnell Douglas and Spain (1983)	1	2	-	5	3	3	1	1	1	3	3
7. General Dynamics and Turkey (1983)	1	3	-	3	3	2	0	2	2	3	2
8. Lockheed and Portugal (TAP) (1983)	1	0	0	0	2	3	0	1	0	1	2
9. Boeing and Saudi Arabia (1984)	1	3	-	3	1	4	0	2	2	1	3
10. Meko and Portugal (1986)	1	1	1	0	0	4	1	2	1	1	2
11. General Dynamics and Greece (1987)	1	1	-	5	3	3	0	2	2	3	2
12. Lockheed and PAF (1987)	0	0	-	0	2	2	0	1	0	0	2
13. Airbus and Portugal (TAP) (1987)	0	3	0	0	2	2	0	2	2	-1	2
14. Westland and Portugal (1990)	0	0	0	0	2	2	1	1	0	0	1
15. Lockheed Martin and South Korea (1992)	1	2	1	1	3	2	2	2	2	3	1
16. USAF and Portugal (1993)	0	1	0	1	2	2	0	2	2	0	2
17. OGMA and Dutch Navy (1997)	1	1	1	0	0	4	2	1	0	1	2

Although on a qualitative basis, the case analysis provides some evidence of some relationships between the variables to be validated under subsequent research where new cases and data may contribute to new inferences with statistical significance.

Combining the quantitative analysis with a qualitative analysis from the cases with higher and lower rate, some conclusions may be taken in order to identify factors to consider when using offsets as an alternative tool to foster innovation among SME's.

4. Conclusions: Use of Offsets as a tool for innovation in the SME's

The analysis carried out under this research shows that, when used by governments as a policy tool to stimulate technological innovation among SME's, Offsets Programs must prioritize the activities before assigning specific players to the offsets projects. Cases 12, 14 and 16 suggest that, taking the opposite approach, the benefits arriving from the projects will remain inside a particular player (in these cases the public company providing maintenance and engineering services to the Portuguese armed forces), with no diffusion on local supply chains where SME's may have been involved.

Priority over the players is also associated with the use of offsets projects in reaction to economic cycles (providing work to big companies when the market is down), or to the integration of new competencies inside non-SME's players, creating new and stronger competition for the local SME's. Under this kind of situation, the *Specialization* effect that takes place inside the particular player may not assure the structural benefits that the Offsets Program must generate when associated to a military or civilian acquisition where significant resources from tax payers are spent.

Offsets Programs should then be activity-focused, rather than player-focused, acting as a tool for the innovation strategies defined for the local industry. Beyond the support of local companies with social and economic importance, these strategies should always consider emerging industries with high degrees of technological intensity, which are more difficult to implement based on the traditional mechanisms of market. Turkey followed a similar approach, trying to allocate a significant share of offsets volume to manage under the *General Dynamics*[xvi] *F-16* fighter Program. A manufacturing facility for this product was installed in the country, responsible for the assembly of most of the fighters acquired by Turkey as well as several aircraft of the same type exported to Egypt.

Under these considerations, the generation of structural benefits to the local industry (including SME's) resulting from the Offsets Program, should be based on the combination of a *Specialization* effect with a *Difusion* effect (figure 4.). SME's participation may either be direct (contractors in the offsets projects), or indirect as local suppliers to the contractors in the projects. For both cases, the creation of new technology-based firms should always be taken into account in order to fulfil gaps in the local supply chains, as well as increasing added value associated to the participation of the local industry in the Program.

The creation of Edisoft by the Portuguese government in 1988 under the Offsets Program related with the purchase of three frigates from the Meko Consortium was an example (case 10). Covering a gap in the market at the time, Edisoft became responsible for the development and maintenance of radar software in the frigates, using some of the absorbed competencies from the offset source (the dutch company Signaal, operating today under the name of Thales Netherlands) to develop innovative products throughout the following decade.

4.1. The Specialization effect under Offsets Program involving SME's

The *Specialization* effect provides the player (big company or SME) with new information that, once absorbed, may contribute to the creation of learning routines acting as a base for the generation of innovative ideas to do *differently* and *better*. This effect may provide the SME's with new technology in material and document forms, and, above all, with non-material technology (access to tacit knowledge) based on the establishment of long-run relationships with best-of-class players acting as sources of external information throughout the implementation of the offsets projects.

The same effect contributes in addressing some of the current challenges in SME's strategies such as (Viana and Hortinha, 1999):

Process and cost in the acquisition of specific knowledge. The lack of human and financial resources devoted to the acquisition of information on products, technologies, markets and competitors places a barrier to the acquisition of specific knowledge by the SME. Financial resources are scarce and people with the required knowledge base spend a significant part of their time dealing with operational issues. Evidence from the analysis shows that, when focused on the creation of long-term business relationships between local SME's and external partners, offsets projects based on continuous information sharing and *learning-by-doing* approaches may decrease the time required for the integration of the competencies in the portfolio of the SME.

Implementation of learning routines. As presented in 2., the theoretical framework considered in the present research considers the theory of institutions and institutional change, where institutions are defined as the *rules of the game*. People learn according to the direction that may provide them with the best individual gains. Regular contact with international players performing best-in-class practices may create an atmosphere in the SME where the rule 'learning and having ideas is worthwhile' may be implemented. Due to the influence of cultural and behavioural aspects in it, this rule falls mainly in the category of the informal constraints, requiring more time to change and reinforcing the long term nature that should assist the implementation of projects under the Offsets Program.

Stronger positioning in new markets. Beyond the acquisition of new knowledge and competencies, offsets projects based on long-term relationships may also provide the SME's with resources to access new markets and consolidate their position. Under a typical eight-year process, preliminary activities in the projects may focus on business intermediation, helping the local SME to enter new markets with existing products. The absorption of new knowledge and technology throughout the project period may provide the SME with some of the conditions to develop new products and services (as well as improving existing processes) tailored to these markets.

Access to niche markets. Due to their dimension and level of resources, the SME's capacity to evaluate opportunities in segments or niches may be very limited. Regular interaction with outside players with more experience in dealing with different markets and niches, as well as a bigger business network, may help the SME to gather and analyze information that may support its decision to explore a new niche market, either based on the existing products and services, or by innovating new solutions.

These benefits may be more significant in industries with higher degrees of technological intensity, depending on their absorptive capacity in selecting informal from external sources and integrating it in new ideas for process optimization or new product/service development. A *multiplying effect* will be associated to the structural benefits generated if the new knowledge and competencies absorbed under the offsets projects may reinforce the existing offer under new business opportunities not directly related with the Offsets Program.

4.2. The Difusion effect under Offsets Program involving SME's

As presented in the previous point, the *Specialization* effect may provide particular companies (SME or not) directly involved in the offsets projects, with the opportunity of doing *differently* and *better*. In order to assure the structural benefits that must be associated to a process resulting from the spending of significant amounts of public resources, a *Difusion* effect must also be considered in order to disseminate the benefits throughout a maximum number of local SME's.

This effect aims to reinforce the link between local SME's through partnering, networking and subcontracting, under the scope of the offsets projects. Beyond the implementation of the offsets projects under a specific programme, the structural benefits may be multiplied if the business started under the Offsets Program remains, generating new opportunities.

Consolidation of supply chains. Offsets projects that may assure a regular flow of business throughout a supply chain for an extended period may contribute to the consolidation of cooperation routines among the SME's taking part in the chain. A planned Program, integrating projects exploring the long-term that should be associated to the implementation of the activities, may provide the ground for that consolidation. This may be achieved through the involvement of the same supply chain in new opportunities not directly related with the Offsets Program, where the level of competitiveness achieved can help in winning new business. At the same time, new business volumes may be generated through different programmes in place, and under the coordination of the local body responsible for their management (Figure 5.).

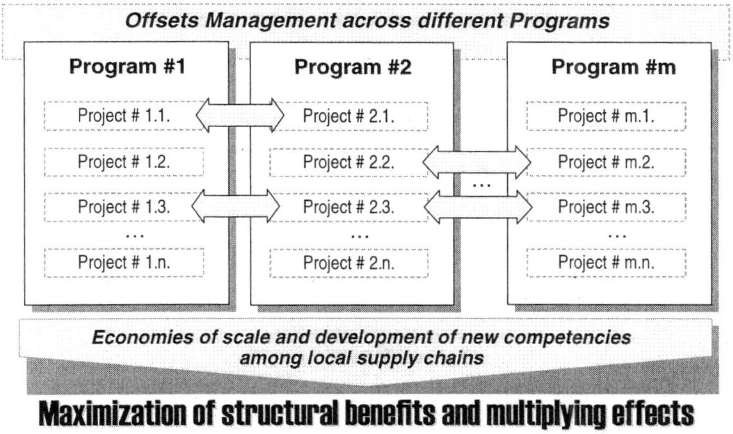

Figure 5. Maximization of structural effects and SME involvement through the coordination of different Programs under development by the same country

Exploitation of common competencies. The pursuit of long-term business relationships able to generate opportunities with stronger added value for the local industry may also provide the necessary framework to explore partnerships and business network scenarios involving SME's, where common competencies may be complemented in order to reach the necessary capacity and economies of scale to implement these opportunities (Doz and Hamel, 1998). Evidence from cases shows the attractiveness of the offsets projects, and they can help in overcoming traditional barriers for cooperation between local competitors. The Portuguese network *HITEC* is an example of this kind of approach, by joining SME's with complementary competencies in software development, electronics and telecommunication technologies.

Resource sharing. As well as common competencies, Offsets Programs may also provide the framework to explore complementarities among the SME's, and to achieve the necessary economies of scale that may contribute to an increase in their level of competitiveness. A proactive attitude aiming to create a competitive offer may help the implementation of the offsets projects, as well as the exploitation of new business in nonlocal markets, based on the new competencies. The search for new partnership opportunities where new competencies can be added in order to reinforce the existing offer may raise the need for additional SME's (that may also be created according to the gaps), helping to maximize the *Difusion* effect, by increasing the number of local SME's involved in the offsets projects. The process for the acquisition of Boeing fighters F/A-18 in 1996 by the Swiss government, established in the offsets agreement that 32 of these aircraft would be assembled by the local company, Emmen Aircraft, owned by the government, which would be responsible for subcontracting local supply chains. According to the Swiss government, this approach would maximize the number of SME's, with more than 300 companies to become directly and indirectly involved in the Program.

This kind of benefit, resulting from the *Difusion* effect, demands a regular interaction between the companies, based on the opportunities that should be provided by the offsets projects. Repetition may play an important role, suggesting that, like *industry selection* for the *Specialization* effect, *volume* is the key factor in assuring an effective *Difusion* effect and maximizing the number of SME's that benefit from the Offsets Program.

4.3. Factors to consider when using offsets to foster technological innovation among SME's

When using offsets as a tool to stimulate innovation and economic growth, local governments should balance the management of the Offsets Program, by considering three groups of industries:

- Industries supplying the base for local employment (independently from the degree of technological intensity);
- Developed industries with high degree of technological intensity;
- Developing or launching of new industries with a high degree of technological intensity.

Offsets should be developed with the public interest as the main priority, suggesting that "What to do" should be defined before "Who is going to implement the activities" (activity-focused). According to the analysis performed on seventeen international cases presented in 3., there is a relation with statistical significance between this orientation and the level of results that may be achieved when offsets are used to change institutions in a direction that may contribute to technological innovation and economic growth.

An activity-focused orientation suggests the existence of innovation strategies by the local government defining the areas of excellence to promote in the local SME's. These priorities may be reflected in the management of the different Offsets Programs where the country is involved. By using this orientation as a guideline, offsets obligors and their local partners may spend less time and resources when defining the projects, the activities of which may generate structural benefits, combining a *Specialization* effect with a *Difusion* effect. This approach may be applied to each of the three groups of industries listed above, where knowledge acquisition by the local companies may help them to find new sources of competitive advantage, even in industries with a lower degree of technological intensity, but playing an important role in the local economic and social structure.

Offsets Programs may also be a powerful tool in bridging gaps in the local industry base, stimulating the start and developing of new ventures, under both the *Specialization* and *Difusion* effects.

New ventures under the *Specialization* effect may take place when the gaps are associated with non-existent or developing industries with a higher degree of technological intensity. Biotechnologies, life sciences, software and materials for the defence industry, and aerospace products and services are some of the examples where new companies may be developed. Cases like the approach to offsets by the United Arab Emirates[xvii] and Saudi

Arabia show how offsets projects may also develop new industries under a symbiotic approach by the local government, offsets obligors and investors.

As presented in 4.2., and under the *Difusion* effect, the volume associated to the offsets projects may provide opportunities for the creation of new technology-based companies able to fulfill gaps in local supply chains. Business generated from the offsets projects may help the sustainability of the venture during the first years, after which the new companies should be able to find other business with strong export potential. A systematic effort in pursuing these opportunities will generate the multiplying effects that must assist the management of Offsets Programs by the local government.

Activities and players' evaluation by the local governmental bodies responsible for the management of the process should consider, for each case, the way the two effects (*Specialization* and *Difusion*) are combined to generate the structural benefits for the local industry. Under an *Activity-focused* approach, players should be seen as the agents to implement the most appropriate activities in order to maximize the structural benefits.

Analysis from the cases with the lower rate under the dependent variable, R, shows that transferring documented technology (licences) was not sufficient. In some of the cases, additional investments by the local government had to be made in order to acquire the material technology needed. Interaction with the *offset source* was limited. The established learning routines did not produce significant results and the creation and consolidation of local supply chains was not achieved, with the *offset recipient* (a big company) doing very little or no subcontracting. The same cases also have in common an option for direct offsets mainly making use of a low developed capability.

At the same time, an analysis of the cases with a higher rate under R shows that activities should consider the total duration of the Offsets Program. In order to innovate, local companies, as *offset recipients*, must change informal constraints, learning to absorb new information and knowledge under the interaction with the *offset source*. And, as stated before, changes in informal constraints is a time-demanding process. Under these considerations, expectations regarding high-value activities should also be managed. Evidence from international cases such the purchase of Northrop's F-5 by the Swiss government in 1975, shows again that R&D projects and product development programmes are very attractive, but are usually difficult to implement in the first years of the offsets projects when informal constraints may still be strong and a trust base is not yet in place (Udis, 1996). Other cases, like Spain with McDonnell Douglas (Molas-Galart, 1996), suggest that such activities should be seen in most of the situations, as a goal to achieve, rather than a demand at the start.

References

[1] Al-Ghrair, A. and Hooper, N. (1996) Saudi Arabia and Offsets, in S. Martin, *The Economics of Offsets – Defence Procurement and Countertrade*, London: Harwood Academic Publishers, pp. 219-244.

[2] Albrecht, K. (2000) *Radar Corporativo,* S.Paulo: Makron Books, pp.137-164.

[3] Ahlstrom, M. and Brege, S., (1998) 'Offset Agreements in Procurement of Large Defence Systems – A General Framework', *14th IMP-Conference Proceedings*, Turku

[4] Confraria, J. (2001) *Desenvolvimento Económico e Política Industrial: A Economia Portuguesa no Process de Integração Europeia,* Lisboa: Universide Católica Editora, pp.69-75.

[5] David, P. A. (2000) 'The Digital Technology Boomerang: New Intellectual Property Rights Threaten Glogal "Open Science"', *Conferência ABCDE (Europe) do Banco Mundial*, Junho 2000.

[6] Doz, Y. L. and Hamel, G. (1998) *Alliance Advantage - The Art of Creating Value through Partnering*, Boston: Harvard Business School Press, pp.245-253.

[7] Egan, C. and Shipley D., (1996) 'Strategic orientations towards countertrade opportunities in emerging markets', *European Journal of Marketing*, Vol. 13, No. 4, pp.102-120.

[8] Féria, L. P. (2000) 'As contrapartidas das Aquisições Militares – Instrumento de Desenvolvimento Económico', *GEPE – Gabinete de Estudos e Prospectiva Económica.*

[9] Fletcher, R., Barrett, N. and Wilkinson, I. F. (1996) 'Countertrade and Internationalisation – An Australian Perspective', *International Commercial Law*, Fevereiro 2000.

[10] Hammond, G. T. (1990) *Countertrade, Offsets and Barter – An International Political Economy,* London: Pinter, pp.6-40.

[11] Hennart, J.-F. (1990) Some Empirical Dimensions of Countertrade, *Journal of International Business Studies*, Second Quarter 1990, pp.243-270.

[12] Kinder, T. and Lancaster, N., 'Absorptive capacity – Building absorptive capacity in a learning region: a socio-technical model', *Science and Public Policy*, Vol. 28, No. 1, February 2001, pp.23-40.

[13] Laranja M., Simões V. C. e Fontes M. (1997) *Inovação Tecnológica - Experiência das empresas portuguesas*, Lisboa: Texto Editora, pp.14-25.

[14] Liesch, P. W. and Palia, A. P. (1999) 'Australian perceptions and experiences of international countertrade with some international comparisons', *European Journal of Marketing*, Vol. 33, No. 5/6, pp.488-511.

[15] Martin, S. (1996) Countertrade and Offsets An Overview of the Theory and Evidence, in S. Martin, *The Economics of Offsets – Defence Procurement and Countertrade*, London: Harwood Academic Publishers, pp. 337-355.

[16] Martin, S. and Hertley, K. (1996) The UK Experience with Offsets, in S. Martin, *The Economics of Offsets – Defence Procurement and Countertrade*, London: Harwood Academic Publishers, pp. 337-355.

[17] Molas-Galart, J. (1996) From Offsets to Industrial Cooperation: Spain's Changing Strategies as an Arms Importer, in S. Martin, *The Economics of Offsets – Defence Procurement and Countertrade*, London: Harwood Academic Publishers, pp.299-371.

[18] Neale, W. N. and Akis, Y. A., Pass, C. L. (1997) 'The life-cycle pattern of east-west countertrade', *European Journal of Marketing*, Vol. 14, No. 4 pp.248-265.

[19] Nelson, R. (1997) 'How new is new growth theory?' *Challenge*, Vol. 40, No. 5, September/October, pp.29-58.

[20] Neto, H. (2002) Uma visão estratégica da economia portuguesa, *Global Economics and Management Review*, Volume VII, n°1 / 2002, pp.9-20.

[21] Neves, J. C. e Rebelo S. (2001) *O Desenvolvimento Económico em Portugal*, Lisboa: Bertrand Editora, pp.95-127.

[22] Nonaka, I. (1997) 'The Knowledge-Creating Company', *Harvard Business Review on Knowledge Management*, Boston: Harvard Business School Press, pp.21-46.

[23] North, D. C. (1991) *Institutions, Institutional Change and Economic Performance*, Boston: Cambridge University Press, pp.3-17.

[24] Pavitt, K. (1984) 'Sectorial patterns of Technical Change: Towards a Taxonomy and a Theory', *Research Policy*, Vol. 13, Nr. 6, pp.343-373.

[25] Redlich, A. and Miscavage, M. (1996) The Business of Offset: A Practitioner's Perspective Case Study: Israel, in S. Martin, *The Economics of Offsets – Defence Procurement and Countertrade*, London: Harwood Academic Publishers, pp. 337-355.

[26] Udis, B. (1996) US-Swiss F-5 Transaction and the Evolution of Swiss Offset Policy, in S. Martin, *The Economics of Offsets – Defence Procurement and Countertrade*, London: Harwood Academic Publishers, pp. 321-335.

[27] Udis, B. (1994) *Offsets in Defense Trade: Costs and Benefits*, report, Department of Economics, University of Colorado, Boulder, May 1994

[28] Udis, B. (2001) *Costs and Benefits of Offsets in Defense Trade*, invited paper presented at Western Economic Association Meetings at San Francisco, July 7, 2001

[29] United States Department of Commerce, 'Offsets in Defense Trade: Fifth Annual Report to Congress', *The DISAM Journal*, Fall 2001, pp.77-107

[30] von Tunzelmann, N. (1999) Big Business growth and decline (review article), *Journal of Economic History*, pp.787-794

[31] Viana, C. e Hortinha, J. (1997) *Marketing Internacional*, Lisboa: Sílabo, pp.397-404.
Welt, L G. B. and Wilson, D. B. (1998) Offsets in the Middle East, *Middle East Policy*, Vol. VI, No.2, www.mepc.org/journal/9810_weltetal.html.

[32] 'As Fragatas Vasco da Gama', Revista da Armada, Abril e Maio, 1991.

[33] 'Countertrade Creates Symbiotic Relationship Between Sellers And Buyers', *BarterNews*, issue #54, www.barternews.com, 2000.

[34] 'Dassault sets up a new venture in Abu Dhabi', *News Room, UAE Offsets Group, www.uaeoffsets.org/pressrelease.htm*, 14 de Abril 1999.

[35] 'EH101 beats off S-92 to meet Portugal's long-range requirement', *Flight International*, Vol. 160, No. 4794, 21-27 Agosto de 2001, pp.22.

[36] 'Greek Military and Civil Offset Programs', Offsets, Industrial Participation and Countertrade, *Presidential Commission on Offsets*, December 4, 2000, www.offsets.brtrc.net/barber2.pdf.

[37] 'Lockheed Martin offers Israel F-16 tie-up', *Flight International*, 5 de Agosto de 2000, Air Transport Intelligence, www.rati.com.

[38] 'Making the transition', *Flight International*, 31 de Março de 1999, Air Transport Intelligence, www.rati.com.

[39] 'Managing Risks In Countertrade Transactions', *BarterNews*, issue #37, www.barternews.com, 1996.

[40] 'Market forces, Approaches to industrial offset and technology transfer are changing in Asian nations'. *Flight International*, Vol. 160, No. 4792, 7-13 Agosto de 2001, pp.38-39.

[41] 'Military Exports: Offsets Demands Continue to Grow', *Report to Congressional Requesters, United States General Accounting Office*, www.clw.org/cat/NSIAD-95-65.pdf, Abril 1996

[42] 'Oasis Signs $195 Million Deal with Gulf Air', *Offsets News – The United Arab Emirates*, Abril 1998, pp.1.

[43] 'Playing the field – Turkish Aerospace Industries is coming of age as it embarks on a bold expansion programmme', *Flight International*, 5 de Dezembro de 2000, Air Transport Intelligence, www.rati.com.

[44] 'Position Paper on Offsets', *European Defense Industries Group*, 1 de Fevereiro de 2001.

[45] 'Pratt & Whitney targets UAE development offsets', *Flight International*, 29 de Novembro de 1995, Air Transport Intelligence, www.rati.com.

[46] 'Switzerland Gets Offsets in F/A 18 Purchase', Offsets, Industrial Participation and Countertrade, *Presidential Commission on Offsets*, December 4, 2000, www.offsets.brtrc.net/barber2.pdf.

[47] 'TAI adopts multi-product strategy to boost orders', *Flight International*, 13 de Junho de 2000, Air Transport Intelligence, www.rati.com.

[48] 'Taiwan's troubles'. *Flight International*, Vol. 160, No. 4792, 7-13 Agosto, pp.32-33.

[49] 'The Netherlands', *American Countertrade Association*, Fevereiro 2000, www.countertrade.org/country/netherlands.htm.

[50] 'The Sixth Rung – Corporate Bartering & Countertrade', *www.countertrade.org*, pp.217-225.

[51] 'Turkey', *American Countertrade Association*, Fevereiro 2000, www.countertrade.org/country/turkey.html.

[52] 'U.S. Government Offset Policy', *Aerospace Industries Association*, Fevereiro 2000, www.aia-aerospace.org/departments/intl/issues/offsets.cfm.

Annex I – Levels of the Independent Variables

X – Focus
0 = Players-focused
1 = Activity-focused

v_1 – *Volume/Frequence of acquisition*
0 = Low
1 = Average
2 = High
3 = Very High

v_2 –Payment/Penalties associated
0 = Independent or no penalties involved
1 = Payment associated or penalties

v_3 – Kind of local investment
0 = Small or non-existent
1 = Financial support per project
2 = Participation in the capital of local companies
3 = *Joint-ventures*
4 = Greenfield
5 = Mixed, according to each project

v_4 – Type of Activities
0 = Purchase / Subcontracting / Intermediation
1 = Intermediation / Subcontracting with transfer of technology
2 = Access to licences
3 = Subcontracting with transfer of technology, and licences

v_5 – Kind of agreements
0 = *Barter*
1 = *Buy-back*
2 = Direct offsets mainly
3 = Balance between direct and indirect offsets
4 = Indirect offsets mainly

v_6 – Level of local competencies
0 = Very low or non-existent
1 = Under development
2 = Developed

v_7 – Social supports towards the acquisition
0 = Not seen as a priority
1 = Important
2 = Strategic

v_8 – Process Visibility
0 = Low
1 = Medium, some references in the press
2 = High visibility, including references from several sources

Endnotes

[i] Paper presented at the NATO Advanced Workshop, '*Defense-related SME's:Analysis and description of current conditions*', Madeira, Portugal, June 2002

[ii] The Prime Contactor is the supplier of the defense equipment, in the form of a single company or a consortium

[iii] The *Offsets Projects* integrating the *Offsets Program*

[iv] Innovation is defined as the process of introducing new products and services the market for which has a strong acceptance and export content

[v] An Offsets Program is defined as the group of projects implemented under the responsibility of the equipment or service supplier as offsets obligor, in order to fulfil its commitment in compensating the local industry

[vi] The *offset source* may be the *offset obligor* or other entity in the circle of influence of the *offset obligor*

[vii] Under the OECD definition, *Technology Intensity* is such as increases with R&D expenditure.

viii The Society of Motor Manufacturing and Traders (in automotive) and the Society of British Aerospace Companies (aerospace) have taken specific activities to disseminate these practices among their members.

[ix] According to Udis (2001), the average duration of Offsets Programs involving US companies as offset obligor is 11 years.

[x] Cases 8., 10., 12., 13. (civil purchase), 14. and 16.

[xi] Case 17. (OGMA and the Dutch Navy)

[xii] Since the effect considered is associated with the transfer of knowledge and technology along supply chains, the term used is *Difusion* from the literature on 'Technology Difusion', instead of *Diffusion.*

[xiii] The practice of buy-back agreements became very usual among sports shoes manufacturers investing in Far Eastern countries

[xiv] Among cases found in the literature, the one with more references and under deeper analysis concerned the acquisition by the British government, in 1986, of Boeing AWACS, instead of the Nimrod solution proposed by the local prime contractor British Aerospace.

[xv]Case 13 concerns the purchase of civil aircraft (Airbus); and Case 17. a contract to perform maintenance services to Lockheed Martin P-3 Orion aircraft.

[xvi] Acquired by Lockheed Martin, operating today as *Lockheed Martin Aerodynamics*

[xvii] In the UAE case, offsets projects were also used to supply social infrastructures

From Downsizing to Modernising Defence in C&E Europe: Opportunities for SME's

Todor TAGAREV
Centre for National Security and Defence Research – Bulgarian Academy of Sciences,
1 "15 Noemvri" Str., Sofia 1040, Bulgaria

Abstract. Defence reforms in most countries in Central and Eastern Europe (CEE) can be characterized by significant downsizing and development of new capabilities. Many of the CEE countries are part of NATO or expect invitations to join and, respectively, to build requisite capabilities. The NATO Defence Capabilities Initiative is one example of required capabilities; it will certainly be refined as a result of September 11, 2001.

Defence reform and modernization in CEE countries may be seen as an effort to acquire, by cooperation or coordination, required defence capabilities and, more broadly, capabilities to protect their national security. That effort is supported through realization of two concepts: role specialization and joint procurement. CEE countries lack experience in implementing these concepts, in particular the latter, in a market environment.

The impact on SME's is examined from two main perspectives: The effect of military downsizing, in particular cases of delayed defence reform with detrimental effect on some defence industries; opportunities created by defence modernization triggered by NATO integration requirements and the necessity to adapt to the new security challenges.

SME's in CEE should exploit the opportunities of defence modernization. Instead of focusing on the need for technology and business management know-how, the paper emphasizes the need for a modern State policy in defining priorities in the development of defence and military capabilities and policy for regulation of the participation of national defence industries, including SME's, in the process of force modernization, as well as on the necessity for national and international cooperation among defence establishments, SME's and the academic sector.

Introduction

A decade after the end of the Cold War the threat of large-scale World War Two-style war has nearly disappeared. The September 11th, 2001 tragedy was followed by unprecedented cooperation in the war on terrorism, involving practically all states in the Euro-Atlantic zone. Today, more than ever, these states recognise their common interests and the need to be able to cooperate effectively in dealing with the security challenges of the new century.

Understanding that the mass armies of the Cold War era are a thing of the past, practically all states in Central and Eastern Europe (CEE) are going through comprehensive defence reforms. Some of the CEE countries started reforms early, others - with a considerable delay. One group of countries deals with the inheritance of large armies, surplus of weapon systems and equipment, rapidly becoming obsolete, and a large defence industrial base with clear specialisation within the Warsaw Pact. A second group of countries, recently regaining independence, builds armed forces almost from scratch and often relies exclusively on foreign supplies and assistance to equip the military.[1]

After briefly describing the background of defence and economic reforms in the CEE, this paper looks at Bulgaria as a typical example of the first group. Bulgaria is undergoing significant *downsizing* of its defence establishment, *modernising* its military to meet the challenges of NATO membership, and *restructuring* its defence industrial base to account for market rules and new political realities.[2] These are complex interwoven processes challenging previous experience and providing opportunities, in particular opportunities for small and medium-sized defence and dual-use enterprises. Although examples from Bulgarian experience are extensively used, most of the tendencies and recommendations equally apply to all CEE countries.

1. Background: Defence and Economic Reforms in CEE countries

Practically all countries in Central and East Europe see their future as members of NATO and the European Union. Although they are at different stages of preparation for accession, it is possible to identify common tendencies that shape the environment for development of defence and dual-use small and medium-sized enterprises.

1.1 Impact of NATO accession

Some of the CEE countries are already NATO members; others expect invitation for membership at the Prague Summit in November 2002, while still others are newcomers or expect to join the Partnership for Peace Programme. Nevertheless, in terms of security, the policies of CEE states are shaped by requirements for *commensurate contribution to the capabilities* of the North Atlantic Alliance to deal with the security challenges of the 21st Century, and these are vastly different from the defence and military capabilities sustained for almost fifty years of antagonism.

Although changes in threat perception are slower than changes in threats to national security, defence planners in NATO, member states and MAP countries [3] already account for the changing security environment. NATO planners implement the *Defence Capabilities Initiative* (DCI), approved at the Washington Summit in 1999. The MAP countries plan to ensure that their forces are being prepared to contribute to the full range of NATO missions. Through the so-called Planning and Review Process (PARP), they develop respective capabilities described in tailored sets of Partnership Goals (PGs). The PGs are roughly equivalent to Force Goals for NATO member countries, and the survey of the aspirants' overall interoperability is virtually identical to the NATO's Defence Planning Questionnaire.[4] Certainly, the Prague Summit will refine the DCI; however, there is a well-elaborated mechanism in place so that the majority of CEE countries, together with NATO member states, plan for development and sustainment of required allied capabilities.

To be able to contribute to the Alliance objectives, the countries in Central and Eastern Europe undergo comprehensive defence reforms. Nevertheless, in planning for required capabilities they start to realise that it is hardly possible—and definitely inefficient—to develop and sustain all necessary capabilities on their own. There is a clear tendency for CEE countries to acquire necessary defence capabilities and more broadly—capabilities to protect national security— *in cooperation or coordination* within the framework of NATO and its Partnership for Peace Programme. Participating to a varying degree in the allied planning process, CEE countries shall nevertheless consider the implementation of two novel concepts: *role specialisation* and *joint procurement.*

Currently, MAP provides for an evolutionary approach to role specialisation. This trend may be accelerated by the new round of NATO enlargement and other decisions of the Prague Summit, so that CEE countries emphasise their traditional strengths and contribute accordingly with national military capabilities. Additionally, the creation of multi-national formations with participation of CEE countries is also gaining momentum.

In regard to joint procurement, however, CEE countries lack the necessary experience and know-how, particularly for joint procurement in a market environment. Furthermore, some countries are in the early phases of developing national defence acquisition and procurement systems adequate to market economies.[5] Yet, due to their participation in the PfP programme,[6] most of the CEE countries are at least informed about defence procurement mechanisms in NATO countries.

Common, or at least compatible, defence acquisition procedures would greatly enhance the opportunities for cooperative, and thus – more efficient, development of required military capabilities. CEE countries may benefit in that respect from the experience of OCCAR and the participating countries.

1.2 Impact of EU accession preparation

The three new NATO members from CEE and seven of the MAP countries,[7] together with Malta and Cyprus, individually hold negotiations for accession into the European Union.

One of the objectives of this pre-accession process is to guarantee that a future EU member country will have a functioning market economy that is competitive within the Union. That includes also acceptance of the EU *aquis* on small and medium-sized enterprises, elaboration of adequate policy for enterprise development and creation of a more favourable business environment for SME's. Given the short recent experience of CEE economies in functioning in market conditions, this is a serious challenge.

The EU enlargement is paralleled by elaboration of the Common European Security and Defence Policy (CESDP) as part of the Common Foreign and Security Policy (CFSP) of the Union. CEE countries, as part of their accession policy, already declared sizeable contributions with military capabilities to the "Headline Goal," formulated during the EU Helsinki Summit. Within the European Union, this process is paralleled by rapid transformation of the European defence industrial base. The European defence industry is rationalised through mergers, creation of joint procurement organisations, i.e., OCCAR, and mechanisms, increased integration of defence research and development (R&D), etc., leading towards a common European armaments policy.[8]

A third area that impacts on the development of defence and dual-use SME's is the participation of CEE countries in the science and technology policy initiatives of the European Union. Most CEE countries are associated in the Fifth and the coming Sixth Framework Programme (FP) of the EU. Within the 6th FP, for example, there is an increased commitment to SME's which will be allocated 15 % of financing for thematic fields in addition to the specific measures for SME's.[9] That makes about € 1.7 billion of the total available budget for research activities. Compared with the 5th FP, the financing of specific measures (cooperative and collective research) for SME's is up by 24 % to € 450 million. In the field of international cooperation, there will be an additional budget for specific actions to match the € 300 million already allocated under research actions.[10]

The interplay of all these factors to provide opportunities and shape the environment for defence and dual-use SME's in CEE is described on the basis of the experience of Bulgaria.

2. From Downsizing to Modernising Bulgarian Armed Forces

From the times of the Warsaw Pact, Bulgaria inherited an disproportionally large defence establishment. The first years of the transition from communism to democracy and from State-run to market economy were particularly difficult. While economic 'growth' was negative and the inflation regularly reached two-digit numbers, the armed forces continued to function by inertia without any thoughtful attempts for adaptation to the post-Cold War era.

1997 marked the beginning of change with the decision of the Bulgarian Government [11] to apply for NATO membership and the adoption of a *National Programme for Preparation for NATO Integration*. Since then Bulgaria has made significant progress in reforming defence. Milestone developments were the adoption of a new Concept for National Security,[12] new Military Doctrine,[13] comprehensive defence reform plans known as "Plan 2004" and an advanced defence resource management system.[14] Developed and implemented under solid civilian scrutiny and in close cooperation with NATO member states, these innovations provided for realistic and affordable development of the Bulgarian armed forces that was also acceptable to society and adequate to the changed strategic environment.

Further, Bulgaria introduced a comprehensive defence planning system that is NATO-compatible, emphasises defence capabilities and provides for flexibility in force development. A simplified version of the defence planning framework is depicted in Figure 1. Rigorous implementation of this planning framework by dedicated professionals provided for balanced development of organisational structures, personnel strength, training levels and investment, all constrained by forecasted levels of defence expenditures. The main quantitative parameters of the defence reform plans, already approved in 1999, are presented in Figure 2, where personnel strength denotes combined military and civilian staff.

Since 1999 several economic forecasts and reform parameters evolved somewhat differently than predicted.[15] Nevertheless, figure 2 gives a qualitatively accurate representation of defence reform and modernisation opportunities.

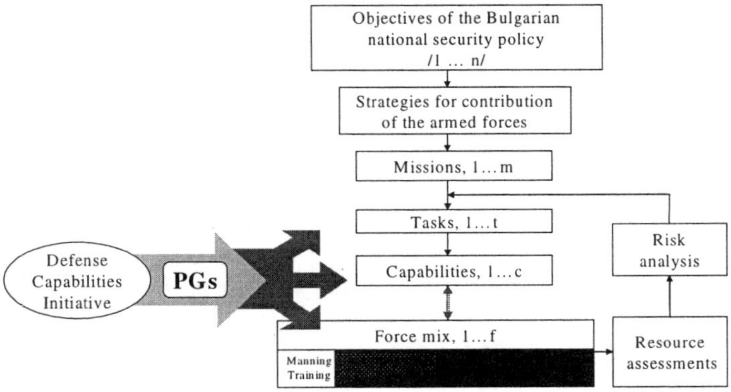

Figure 1: Framework for capability-based defence and force planning

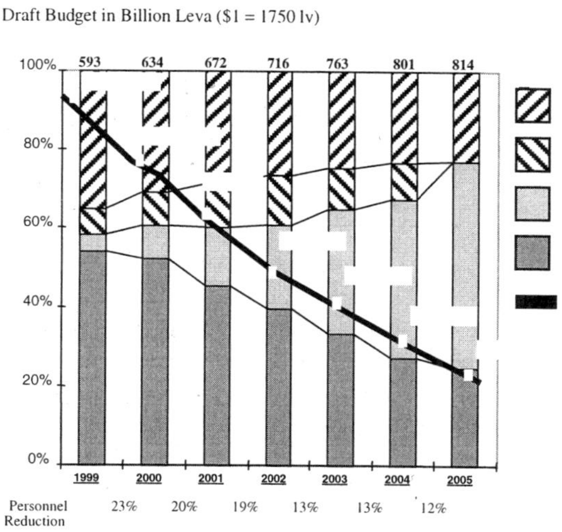

Figure 2: Major Bulgarian Defence Reform Parameters, 1999-2005

At the start of defence reform, c. 1998, only 3-4 percent of the defence budget was spent on investment, and training levels reached a decade-long minimum. Today, defence reform and the related budget distribution provide for the necessary maintenance of the planned force structure, training according to NATO standards and modernising the force to build and sustain the capabilities necessary for national defence, a reasonable contribution to NATO or EU crisis response operations and a significant contribution to collective defence.

Strictly following reform plans, after 2004 the modernisation budget [16] will amount to 25 percent of the Bulgarian defence budget. Thus, Bulgaria will meet the Common European Security and Defence Policy target figure [17] and will exceed the NATO floor for modernisation spending. [18]

Recently, senior military leaders claimed that it is realistic to allot 30 percent of the defence budget in the 'near years' for 'modernisation of weapon systems and equipment' and to complete the modernisation of the Bulgarian Armed Forces by the year 2010.[19] That statement was substantiated by claims that after 2007 at least 35 percent of the defence budget would be spent on investment.[20] However, given the updated defence reform "Plan 2004,"[21] current economic forecasts and the expected level of defence expenditures, the prospects are that 30 percent of the defence budget will be allotted to 'capital expenditures,' which is equivalent to spending approximately a quarter of the defence budget on investments.

The increase in defence investments is paralleled by increasing levels of training, introduction of new principles of defence management, and preparation for NATO and EU membership, which in combination provide a favourable environment for the development of defence and dual-use SME's in Bulgaria.

3. SMEs within the Bulgarian Defence Industry

As a member of the Warsaw Pact, Bulgaria developed a large defence industrial complex specialising in small arms and light weapons, ammunitions, radio, radio-relay and navigation equipment, certain types of radars, computing hardware and software, and overhaul of all major weapon platforms. Direct employment in the defence industry during the 1980's was estimated at 115,000 persons, with additional 400,000 persons in direct support to defence industries. At the peak in the late 1980's, the national defence industry produced nearly 10 percent of the country's GDP and the exports exceeded 1 billion USD annually.[22] All production facilities were largely export-oriented. The Bulgarian armed forces consumed approximately 5 percent of the production; between 20 and 30 percent was intended for the armed forces of other Warsaw Pact countries, while the rest was exported to customers in the Middle East, Asia, Africa and Latin America.[23]

The defence industry was run entirely by the State as part of the centralised planned economy. The Ministry of Defence itself managed a considerable industrial capacity, assigned mostly for maintenance and overhaul of weapon systems and military equipment. Market mechanisms were largely unfamiliar, and there was no particular treatment of small and medium enterprises.

The first years of the transition had a detrimental effect on most defence industries. Since the defence expenditures as a percentage of the GDP went down, and defence reform was delayed, the Bulgarian armed forces practically disappeared as a customer of the national defence industry, consuming in the mid 1990's less than 1 percent of its production.[24] Furthermore, Bulgaria strictly adhered to decisions of the international community on imposition of sanctions and embargos, and the State enacted strict laws for control of arms export. As a result, the defence companies lost many of their traditional customers.

Currently, with the ongoing defence reform and the policy for integration in NATO and the EU, Bulgaria provides a significantly better environment for defence industries, and SME's in particular.

The development of SME's is strongly encouraged by the European Union and is the subject of pre-accession negotiations. In 2000, Bulgaria provisionally closed Chapter 16 of the negotiations on "*Small and medium-sized enterprises.*"[25] Efforts continued to create a more favourable business environment, and some of the reforms will benefit small enterprises in particular. An inter-institutional working group proposed simplification or abolition of 121 licensing and other requirements. At the beginning of 2001, a number of tax reforms were implemented to encourage incentives and business development, including a 5-percentage point reduction in the rate of profit tax and a reduction of the VAT reclaim period. Dialogue between the public administration and the business community intensified, with the Ministry of Economy holding regular consultative meetings with business associations. A recently created *National SME Agency* encouraged several new initiatives. There is still lack of venture capital for the provision of external equity finance to small businesses, but the State-owned Promotional Bank increased its lending activities to SME's, while a variety of other State-owned and commercial banks stepped up their SME lending, some with the support of donor credit lines.[26]

According to the estimate of the European Commission, Bulgaria's SME policy is in line with the principles and objectives of EC enterprise policy. However, the Bulgarian definition of SME's, notably as regards medium-sized enterprises, still differs from the EC definition and should be aligned; the country needs to intensify its efforts to develop a more favourable business environment, and to address the administrative, financial, fiscal and managerial obstacles to SME creation and development. The EU further recommends developing the capacity of business associations and ensuring they are consulted in an even-handed and transparent manner. According to recent estimates, in 2001 25 percent of the value added and 40 percent of the jobs in Bulgaria were created by small and medium-sized enterprises.[27]

Currently, the defence industry of Bulgaria directly employs some 25,000 people. Among 77 companies with defence or dual-use production capacity, 5.7 percent are classified as *small*, 63.6 percent as *medium-sized*,[28] and 30.7 percent as *large*.[29] With a few exceptions, the defence companies were privatised. Often, though, companies were sold to existing management,[30] and the 'privatisation' did not attract strategic investors. The industry often lacks modern management, marketing, and financial skills.[31]

Nevertheless, the industry is gradually restructuring. Production is being oriented along the requirements of NATO standards. The companies seek new markets and new Western partners. Among the strategic assets and skills of the Bulgarian defence industry, foreign observers often point to the availability of skilled, low cost personnel with significant engineering capabilities, modern and flexible dual-use industrial capabilities, and significant maintenance and repair capabilities.[32]

The start of modernisation of the Bulgarian armed forces creates considerable opportunities for the defence and dual-use SME's. Whether these opportunities will be used depends on how the State and the companies will meet a number of challenges.

4. Challenges for the Development of Defence and Dual-use SME's

Defence modernisation and increasing training levels, triggered by NATO integration requirements and the necessity to adapt to the new security challenges, create opportunities for SME's. The development of SME's is also facilitated by the policies for EU integration. However, while market rules are important, defence is never an open market. Therefore, the State and business, in a joint effort, need to promote a favourable environment for SME's, adhering to the principles of outsourcing, transparency and partnership. Another challenge is to develop and sustain the respective knowledge infrastructure.

4.1 Outsourcing

Focusing on military capabilities while downsizing, the Bulgarian Ministry of Defence is strongly encouraged to outsource many of its activities, and all those activities that are "not essentially military." The Bulgarian Parliament defined outsourcing in such cases as a legislative requirement to defence management.[33] That provides opportunities for SME's not only in supplying equipment, but also in maintenance and repair, infrastructure support, providing "assets from trade" in crisis situations, etc. Particular attention is deserved by the outsourcing of defence research and development (R&D), and the cooperation among SME's and academia in R&D, testing and evaluation, advanced technology demonstrations, concept development and experimentation, etc.

4.2 Transparency

The strategic management of SME's is highly dependent on the transparency of defence policy and in particular the transparency of defence acquisition. It is important that procurement tenders are conducted in a transparent manner and reference to secrecy of procurement information is strictly limited.[34] It is no less important that the government informs the business community of its long-term plans for defence reform and force modernisation [35] and provides sufficient information so that an informed person would be able to relate mission needs (capability gaps) to user/operational requirements to system and technical requirements to procurement decisions.[36]

Another aspect of transparency relates to the licensing regimes that make enterprise creation difficult and, in particular, the functioning of SME's. The remaining licensing requirements still impose a significant burden on enterprises in terms of time and money. Their non-transparent nature provides an opportunity for the public administration to use their discretionary power, leaving open opportunities for corruption.[37]

Business associations of defence and dual-use SME's may contribute to the pressure on governments to promote transparency, thus facilitating the development of a favourable business environment.

4.3 Strategic partnerships

Force modernisation efforts are heavily influenced by requirements of NATO integration. Particularly important is the requirement to provide *interoperability* of command and control, communications and information systems, air and sea surveillance, logistics and other host-nation support capabilities, major weapon systems and equipment. As a rule, SME's in CEE countries can hardly provide interoperable solutions that are at the same time efficient and cost-effective. Such solutions in key procurement cases may be provided by major Western defence contractors; however national interests call for participation of the local industry. Therefore, instead of applying trade-distorting offset requirements, it is advisable that CEE countries use major procurement cases to establish long-term strategic partnerships with 'first-tier' defence contractors.[38]

However, in order to provide a really long-term, mutually beneficial and market-driven partnership, CEE countries need to have competitive advantages in specific niches. Since the low-cost labour does not provide long-term advantage, these 'towers of excellence' are crucial for the international position of a country.[39] In that respect, the national effort to sustain requisite knowledge infrastructure is indispensable.

4.4 Knowledge Infrastructure

One major challenge to SME's in CEE countries is the access to high-tech innovative solutions to defence needs that may be provided through enhanced coordination at a national level or through international cooperation. In any case, each CEE country needs a framework strategy for the development of research capacity and knowledge infrastructure that accounts for the trend of increasing internationalisation of science, including defence science and technology.[40]

WEAG and NATO provide opportunities for such cooperation. The European Union provides additional possibilities through its Fifth and the coming Sixth Framework Programme with which most CEE countries are fully associated. In regard to Bulgaria, the EU assessment is that

> For the further development of the sector and for an effective integration of Bulgaria into the European Research Area, it is important to increase the gross domestic expenditure on research and technological development. The promotion of science and research in Bulgaria will particularly need a further increase of business expenditure on research and development (BERD), i.e. of expenditure on science and research activities from the industry, small and medium-sized enterprises and the private business sector in general. Furthermore, there is a need to foster research activities in universities and to introduce proper risk capital legislation. Emigration of young and experienced specialists continues to hamper the development of this sector.

Other CEE countries face similar problems. According to the Bulgarian experience, successful defence SME's that attract high interest of foreign counterparts and major defence contractors, are those that combine modern managerial practices and production efficiency with considerable investment in R&D.[41]

Conclusion

During significant downsizing of their armed forces, the countries in Central and Eastern Europe are launching comprehensive force modernisation programmes that provide opportunities for developing strong sectors of defence and dual-use small and medium enterprises. To use these opportunities, SME's need business management know-how and technology. Furthermore, even though the role of the State is diminishing, there is a need for a modern state/national policy in defining priorities in the development of defence and military capabilities and a policy for regulation of the participation of national defence industries, including SME's, in the process of force modernisation. Essential components of such state policy are (1) the establishment of national cooperation among government, SME's and academia in order to sustain niche competitive advantages and more comprehensive knowledge infrastructure; and (2) the encouragement of international cooperation of SME's, in particular the development of strategic partnerships with the world leaders in the defence industry. It is expected that international cooperation will be facilitated by introduction of compatible defence acquisition systems and will receive a boost with the increase in specialisation of CEE countries in providing military capabilities.

ABOUT THE AUTHOR: **Todor TAGAREV** is Director of Programmes of the Centre for National Security and Defence Research at the Bulgarian Academy of Sciences. He was the first Director of the Defence Planning Directorate of the Bulgarian Ministry of Defence since its establishment in early 1999. From May 2001 until late 2001, he served as Director for Armaments Policy and National Armaments Director. He graduated from the Bulgarian Air Force Academy in 1982 with an M.Sc. degree in electrical engineering and received a PhD degree in systems and control from Zhukovsky Air Force Engineering Academy, Moscow, in 1989. Dr. Tagarev is a 1994 Distinguished Graduate of the US Air Command and Staff College at Maxwell Air Force Base, Ala., and a 1994 Distinguished Young AFCEAN. He is Editor of *Information & Security: An International Journal*. Dr. Tagarev specialises in the integration of information technology with security and defence policy, computer modelling and prediction of complex processes. *E-mail*: tagarev@space.bas.bg.

Endnotes

1 Bulgaria, the Czech Republic, Hungary, Poland, Romania and Slovakia are in the first group. Estonia, Latvia and Lithuania are typical representatives of the second group. There is no attempt in this paper to deal with the other successor states of the Soviet Union.

2 For a discussion on similarities in defence reforms in CEE countries the reader may refer to *Rebuilding the Armed Forces for the XXI-st Century* (Bucharest: Institute for Political Studies of Defence and Military History, 1999).

3 The countries actively seeking NATO membership through implementation of tailored Membership Action Plans (MAP). Currently, these are Albania, Bulgaria, Estonia, Latvia, Lithuania, Macedonia, Romania, Slovakia and Slovenia. In 2002 Croatia also joined the MAP process.

[4] Franc Boland, "Mapping the Future," *NATO Review* (Spring 2002), <http://www.nato.int/docu/review/2002/issue1/military.html> (27 May 2002).

[5] For the status of defence procurement procedures in the countries of South Eastern Europe the reader may refer to Todor Tagarev, ed., *Transparency in Defence Policy, Military Budgeting and Procurement* (Sofia: Geneva Centre for the Democratic Control of Armed Forces and George C. Marshall – Bulgaria, 2002).

[6] And to an extent – in the activity of WEAG.

[7] Bulgaria, Estonia, Latvia, Lithuania, Romania, Slovakia and Slovenia.

[8] However, although CFSP is one of the areas of pre-accession negotiations, to the best of the author's knowledge CEE countries have not been involved in debates on the development of the European defence industrial base.

[9] Philippe Busquin, "European research steers a new course," *RTD Magazine for European Research* 33 (April 2002): 3-5, <http://europa.eu.int/comm/research/rtdinfo/pdf/rtd33_en.pdf> (22 May 2002).

[10] "The home straight: A preview of the Sixth Framework Programme due to be adopted this summer," *RTD Magazine for European Research* 33 (April 2002): 6-8, <http://europa.eu.int/comm/research/rtdinfo/pdf/rtd33_en.pdf> (22 May 2002).

[11] In February 1997 the decision was made by the caretaker government appointed by President Petar Stoyanov. This decision was reaffirmed by the National Assembly (the Bulgarian one-chamber Parliament) on 8 May 1997 with the *National Consensus Declaration.*

[12] National Security Concept of the Republic of Bulgaria, *State Gazette* 46 (22 April 1998). Available full text in English at http://www.md.government.bg.

[13] *Military Doctrine of the Republic of Bulgaria*, Approved by the XXXVIII National Assembly of the Republic of Bulgaria on April 8, 1999 (Sofia: Military Publishing House, 1999). Full text in English is available at http://www.md.government.bg.

[14] See for example Dobromir Totev, "Bulgarian Defence Resource Management System - Vehicle for Transparency in Defence Planning and Budgeting," in *Transparency in Defence Policy, Military Budgeting and Procurement*, ed. Todor Tagarev (Sofia: Geneva Centre for the Democratic Control of Armed Forces and George C. Marshall - Bulgaria, 2002), 71-84.

[15] More recent figures are provided in Todor Tagarev, "Prerequisites and Approaches to Force Modernization in a Transition Period," *Information & Security: An International Journal* 6 (2001): 30-52, <http://www.isn.ethz.ch/onlinepubli/publihouse/infosecurity/volume_6/f4/f4_index.htm> (12 March 2002).

[16] Definitions for "modernisation" and "investment" differ. According to Bulgarian legislation, modernisation, re-equipment and new defence infrastructure spending constitutes only part of the 'capital investment' budget. The latter further includes overhaul of weapon systems, equipment and infrastructure, as well as building of barracks, housing, etc. The adoption of common NATO definitions is strongly recommended.

[17] Gilles Andreani, Christoph Bertram and Charles Grant, *Europe's Military Revolution* (London: Centre for European Reform, March 2001).

[18] Thomas S. Szayna, *NATO Enlargement, 2000-2015: Determinants and Implications for Defense Planning and Shaping*, RAND Report MR-1243-AF (RAND Corporation, 2001), <http://www.rand.org/publications/MR/MR1243/> (7 May 2001).

[19] Miho Mihov, Chief of the General Staff of the Bulgarian Armed Forces, "Modernization Timelines Shall Be Shortened by Five Years," Mediapool.BG (2 June 2002) <http://www.mediapool.bg/site/bulgaria/2002/06/02/0002.shtml>.

[20] Statement by Major General Sabi Sabev, Chief of J5 at the General Staff of the Bulgarian Armed Forces, in a presentation to the 2002 Spring Session of the NATO Parliamentary Assembly (Sofia, 24-28 May 2002).

[21] In December 2001 the Council of Ministers (the Government) with Decree # 301 approved amendments to the "*Plan for Organisational Development of the Armed Forces of the Republic of Bulgaria till the Year 2004.*" The initial version of "Plan 2004" was approved with Decree # 200, November 1999, *State Gazette*, no. 98 (1999).

[22] Curtis M. Coward and Jeffrey B. Bialos, *The Bulgarian Defense Industry: Strategic Options for Transformation, Reorientation and NATO Integration* (Washington: The Atlantic Council of the United States, July 2001), <http://www.acus.org/Publications/policypapers/transatlanticrelations/bulgaria.pdf> (27 July 2001).

[23] Valeri Ratchev and Ivan Kolev, "Towards a New Defense Industrial Base," in *Bulgaria for NATO 2002*, ed. Ognyan Minchev, Valeri Ratchev and Marin Lessenski (Sofia: Institute for Regional and International Studies, 2002), 279-294. <www.iris-bg.org/pub.html#b11> (27 May 2002).

[24] Ratchev and Kolev, "Towards a New Defense Industrial Base."

[25] *Accession Negotiations: State of Play* (European Commission - Enlargement DG, April 2002), <http://europa.eu.int/comm/enlargement/negotiations/pdf/stateofplay_april2002.pdf> (27 May 2002).

[26] The conclusions in this and the following paragraph are due to the *2001 Regular Report on Bulgaria's Progress towards Accession* (Brussels: Commission of the European Communities, 13 November 2001).

[27] Roumen Vesselinov, "An Emerging Economic Factor," in *Bulgaria for NATO 2002*, ed. Ognyan Minchev, Valeri Ratchev and Marin Lessenski (Sofia: Institute for Regional and International Studies, 2002), 61-59. <www.iris-bg.org/pub.html#b11> (27 May 2002).

[28] However, Bulgaria considers as 'medium-sized' companies with 51-500 employees, which differs from the EU definition of maximum 250 employees, maximum turnover of 40 million Euro or maximum balance sheet-total of 27 million Euro, and independence (which means that less than 25 % are owned by one enterprise, or jointly by several enterprises, falling outside the definition of an SME).

[29] Ratchev and Kolev, "Towards a New Defense Industrial Base."

[30] Formally – to specially created Workers-Management Organisations.

[31] Coward and Bialos, *The Bulgarian Defense Industry*, p. viii.

[32] See for example Coward and Bialos, *The Bulgarian Defense Industry*.

[33] *Military Doctrine of the Republic of Bulgaria*, Article 98. However, no explicit procedures have been established for enforcing that requirement.

[34] Recent experience shows, though, that the administration is too easily tempted to refer to secrecy in defence procurement cases.

[35] Such attempt was made in Bulgaria with the preparation of a comprehensive Force Modernisation Plan. For details refer to Tagarev, "Prerequisites and Approaches to Force Modernization."

[36] The current status is presented in detail in Todor Tagarev, ed., *Transparency in Defence Policy, Military Budgeting and Procurement.*

[37] *Accession Negotiations: State of Play*, p. 32.

[38] See for example Wally Struys, "Leveraging Defense Procurement: The Belgian Offset Policy," presented at *Defense Economics III. Defense Industry and the State* (Wildbad Kreuth, Germany: George C. Marshall European Center for Security Studies, 26-30 August 2001).

[39] E.I.L.D.G. Margherita, "Defence Research and Technology in the Netherlands," presented at the *First International Conference on Defence Research and Technology* (Plovdiv, Bulgaria: 30 May – 1 June 2002).

[40] For the basics of the Bulgarian strategy the reader may refer to Todor Tagarev, "International R&D Co-operation as a Driver for Re-engineering Bulgarian Defense Research," presentation at the *Joint RTB/Partner Session* (The Hague, 22 March 2001), <www.icsr.bas.bg/cnsdr/Documents_files/Tagarev_RTB2001.ppt> (12 Nov 2001).

[41] One example is the company Electron Progress JSC with 200 employees, <www.bgnet.bg/~eprogres>.

Defence Related SME's
F.D. Carvalho (Ed.)
IOS Press, 2004

Defence-Related SME's: Analysis and Description of Current Conditions; The Challenge of Re-Adjustment in the 21st Century

Professor Trevor TAYLOR[1]
Cranfield University, United Kingdom

Abstract. The paper examines the implications of three key developments for SME's in defence. The first is the trend on the supply side towards the creation of transnational defence companies that dominate prime contracting in defence. The second is the increasing acceptance of the key elements of Smart Acquisition, not least by OCCAR. Increasingly governments are looking to companies that will respond to ITTS using functional requirements and will commit themselves to the fixed price development and production contracts. Governments are ready to operate in partner relationships with successful primes. The third is the increased frequency of military operations by European states and a return to concerns about industrial support for such operations. Such concerns largely disappeared during the deterrence dominated years of the Cold War but have reappeared on the agenda since 1990.

The paper argues that, if SME's are not to be weakened by these processes, behavioural changes are needed from procurement bodies, prime contractors and SME's themselves. It offers suggestions for some changes to sustain the flexibility and innovation attributes that SME's can bring to defence.

Introduction

The supply chains associated with contemporary defence platforms and even individual weapons systems are extensive and complex, and becoming more so. Unless a major and sustained effort is made, the supply chains generating a platform or a weapon will not be understood by any organisation[2].

Many Western governments are keen that SME's should have a significant role in defence acquisition because such firms are often innovative and flexible, lacking as they do the bureaucratic and formal structures of larger companies[3]. It may also be thought that defence contracts can be a means of boosting the turnover of such small firms so that they can become larger and better established, serving the civilian market as well. Small firms, however, have particular problems in raising capital, may be subject to intense competitive forces unless they have a specific niche of valued expertise, and usually lack the resources to undertake research[4]. Defence prime contracting has, of course, become dominated by a handful of large firms that are mainly based in the United States and Western Europe.

This paper analyses the prospects for SME's in defence acquisition and suggests measures that could strengthen their position by analysing the factors that strengthen or weaken the outlook for SME's on the supply side and the demand side. The supply side relates to the practices associated with the defence industry, especially the sub-contracting activities of larger commercial organisations. The demand side refers to the acquisition behaviour of governments. The paper essentially populates Figure 1 with some arguments and contemporary trends.

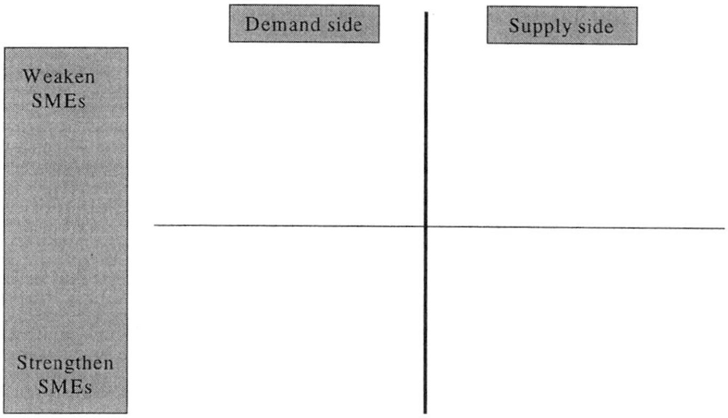

Figure 1: Factors affecting SME's in defence acquisition

1. The demand side

Some governments, and especially the United Kingdom, have put themselves in a poor position to support SME's because of the procurement strategies they have adopted. There are several elements of British procurement and, more widely, acquisition policy that are of note.

1.1 Prime Contracting

The UK has sought whenever possible to appoint a prime contractor for major projects which will take full contractual responsibility for the development and delivery of equipment on time, on cost and to the specified quality. This makes the Government reluctant to specify particular sub-contractors (or types of sub-contractors) that should be used or to provide much in the way of government-furnished equipment. Any problems with such elements in the supply chain may then be used by the prime contractor to justify any shortfalls in the overall performance of the system.

The US is slightly different: in the US there is again a clear preference for the specified prime contractor, but the government there appears less concerned about constraining the prime contractor's freedom of action. This is in part because the US is often less concerned about cost or time overruns than is the UK and the US remains content often to use cost-plus contracts for development work. Overall the US equipment budget is under less pressure than that of the UK. The obvious implication is that, the more a government attempts to build significant defence capability from a very limited budget, as is the case in the UK, the less attention is likely to be paid to considerations such as the role of SME's in the supply chain.

Prime contracting as a practice generally encourages the emergence of defence industrial giants because smaller firms lack the resources to be able to remedy a problematic commitment. The former GEC company paid an unknown amount from its own funds to complete the development of the Phoenix reconnaissance drone once the money specified by government for the system had been exhausted. A smaller firm, including the firm responsible for the engine, could not have done this. More than a decade ago Westland Helicopters realised it was not financially credible as a prime contractor for the anti-submarine version of the EH.101 in the light of the technical risks involved. It thus turned to IBM with its electronics expertise and financial reserves. IBM became the prime contractor with Westland-Agusta supplying the airframe that had been developed on a collaborative basis. Today BAE Systems is understood to be carrying a significant financial burden because the development and production of the latest anti-submarine version of the Nimrod aircraft is taking a longer period than provided for in the contract.

1.2 Function-based contracting

The United Kingdom has moved far away from the procurement practices of the Royal Navy as late as the 1970's. The Royal Navy's architects essentially used to design the ships and then get industry to build them. The UK today seeks to minimise the technical specifications in its contracts and to contract instead for functional performance. The idea is to allow the prime contractor maximum freedom to respond to the performance challenges set by the MoD. However, such contracts clearly make it harder for the MoD to influence the precise content of any system. If a major company decides to generate a product that has little content from SME's, the Defence Procurement Agency can scarcely complain if the product meets the requirement. Functional contracting is also a feature of American defence procurement, reflecting the basic consideration that the armed forces should focus their attention on what they want to be able to do, the capability needed. Industry should be expert in seeing the optimum solution to provide that capability. Governments still need the skills to assess the technical merit and risk of any proposal, but that is not the same as taking responsibility for the design of equipment.

Intriguingly, in parts of the commercial world, the reverse trend can be discerned in which the customer takes responsibility for the design of a desired product, with help from the supplier. This is occurring especially when a customer is unsure of what precisely it wants and when it may have trouble articulating its desires.[5]

1.3 Fixed and firm price contracting

As has already been indicated, the UK has moved as far as it can from cost-plus contracting and instead favours fixed or firm price contracts, even for development work when possible. In cost-plus contracting a Ministry needs to penetrate the main contractor and much of the supply chain to check that over-charging is not occurring. If the MoD agrees a fixed price, especially after a competition has been held, it is harder to justify the allocation of MoD staff to get involved in examining the supply chain and the cost structure of the prime contractor. With intense pressures on the defence budget, the UK as a consequence has a relatively small number of people involved in defence procurement, at least compared with the United States. In the US the procurement authorities, dominated by the individual armed services, still can deploy significant staff looking into the practices of its suppliers.

1.4 Competitive contracting

A further feature of UK procurement strategy and arguably a growing element in the rest of Europe is an emphasis on competitive contracting. The impact of competitive contracting on SME's is multi-dimensional.

Competitive contracting should benefit the smaller independent firms that are competing against the in-house facilities of the giant defence companies. Competing giants have to minimise their costs if they are to prevail in competitions and should be ready to outsource those areas where their own factories are not competitive. If companies do bid for prime contractorships on the basis of allocating work to their own (uncompetitive) plants, the chances rise that they will either not win the contract or lose money on its implementation. Equally, the large primes are also in a position to make strategic decisions about transfer prices within the company.

On the other hand, competitive contracting rewards those firms that have the resources to fund the bidding process itself. Large firms find this easier than small ones. The costs of bidding are obviously even higher when governments or prime contractors expect smaller firms, at their own expense, to undertake significant development work as part of the bidding process.

Also, there is the intellectual property aspect. Small companies that come up with an innovative idea or technology face the issue of whether they share their thoughts with a potential customer. In public procurement generally, and not just the defence sector, there is strong stress on the value of competitive contracting. Such practices are after all a significant factor hindering corrupt practices. Nevertheless, there is little incentive for the small firm that has seen an innovative way to generate a capability to share its thought with a potential customer if it feels the customer will then issue an Invitation to Tender that could at least signal how the capability might be developed. Ministries of defence may still be tempted to put companies' intellectual property, provided to support a possible sale, into the public domain in order to generate a competitive process[6].

1.5 Integrated Project Teams

Integrated Project Teams (IPTs), bringing together experts with different specialisms under a specified leader, have become important features of defence procurement in the US and the UK. Such teams can and often do include representatives from the industry supplying goods or services. However, it is already clear in the UK that big firms can more easily afford the resources to keep continuous contact with IPTs. Moreover, when IPTs are used in conjunction with an emphasis on prime contracting, the IPT leader may become reluctant to speak to firms lower in the supply chain for fear of undermining the position of the prime contractor. Prime contractors are unlikely to relish the prospect of firms considerably below them in the supply chain perhaps presenting radical ideas to the IPT customer. Overall IPTs can become difficult forums for SME's.

1.6. The Demand for Systems of Systems

The particular contribution of SME's often comes from innovative individual technologies and devices or new manufacturing processes. However, the governments of today's advanced states are less interested in individual breakthroughs and more focused on the benefits that arise from the integration of sensors, communication systems, command and control mechanisms, platforms and weapons. These are the developments must closely associated with the next, much-anticipated Revolution in Military Affairs. Governments are looking particularly to industrial systems integrators to provide them with what they want. While small firms can sometimes generate niche expertise in the integration of information systems in particular, for the most part systems integration is for the large and indeed broad-based firms that have expert knowledge of a range of technologies and how they interact.

1.7 Moving away from 'juste retour' in collaborative projects

Historically there have been many issues about the roles of SME's in collaborative projects that have been governed by 'juste retour' work share arrangements. It is unlikely that many British companies would like to see a repeat of the Eurofighter exercise where 'juste retour' percentages were applied, not just to the airframe, the engine and the radar, but to many sub-systems below that. With governments insistent that competition should be used wherever possible for the selection of sub-contractors, the result was that much time and expense was needed for the creation of competing teams of sub-contractors. Firms wishing to bid had to find suitable partners and then agree on a suitable division of the work. There has to be a better way. According to the ambitions of OCCAR and (perhaps) its sponsoring states, 'juste retour' is not to be applied to individual projects in future. At least it should be used in a less rigid form than in the past. This could generate a different pattern with some SME's gaining from being able to secure a contract for the whole of a sub-sub-system while others miss out entirely.

2. Supply side developments

2.1 Prime contractor re-structuring across national borders

Moving to the industrial side of the account, it is clear that the defence industrial re-structuring that accelerated in the latter half of the 1990's threatened many relationships between prime contractors and their suppliers. Not only has the number of major defence firms been much reduced, many of the new organisations have a transnational character. It is apparent that firms such as European Helicopter Industries, Eurocopter, EADS, Thales, Matra-BAe Dynamics-Alenia MBDA) and BAE Systems need to organise themselves and their supply chains on a scale above the national if they are to have a chance of competing with the large US firms serving the US market. This is of clear importance for the smaller firms accustomed to serving national prime contractors.

Overall, changes in collaborative procurement practices (demand-side) seem to be less important in the medium term than the move to multinational defence companies in Europe. Particularly associated with Thales is the idea that these companies will be 'multi-domestic' i.e. they will seek to present an image of being a national firm in all the countries in which they operate. Part of any national image is likely to include use of a supply chain with a significant national content. This is re-assuring news for SME's. However, the more that such companies come under pressure from governments to minimise the costs of the defence equipment that they develop and produce, the more companies will want to rationalise their supply chains so that they comprise only the most effective and efficient suppliers. Also, there is a regulatory dimension relating to governmental behaviour. The more that European and even NATO governments liberalise controls on the international transfer of defence-related technologies, as per the 1998 Letter of Intent and the Framework Agreement of 2000, the more will larger companies be drawn to adopt multinational rather than national supply chains.

2.2 The fashion for outsourcing

An important factor boosting the prospects of SMEs is the widespread fashion supporting outsourcing[7]. This affects the civilian as well as the defence world. The logic is that companies should focus on those activities where they see themselves as being internationally or even globally competitive and/or those activities that are vital for the future success of their business. All other activities can be outsourced to specialist businesses that can provide a superior product or service at lower cost. Implementation of this logic has meant that vertically integrated firms have become rare and, in the automobile-manufacturing sector, about 70% of the value of the final product is provided by sub-contractors. The same percentage probably applies in most areas of defence manufacturing and the result is to offer real opportunities to small firms with niche but valued capabilities. Some analysts in Sweden expect defence outsourcing to proceed further in that country, but not necessarily to the benefit of Swedish firms.[8] An important implication of extensive outsourcing is that the management of suppliers and relationships with them becomes a key competence of the lead firm.

2.3 Appeal of the civilian sector

Until the late 1990's the defence sector itself must have appeared unappealing for many SME's in the West because of its (at best) limited growth. In contrast, the civilian computing and telecommunications areas, as well as aerospace, were doing well. Then the telecommunications bubble burst, but generally Western economies are continuing to grow faster than does the defence sector. It is not clear how defence can hope to attract the attention of the brightest of the SME's that will be looking for rapid expansion. Clearly those SME's that are tied to the defence sector and find it difficult to move to serving the civilian market will be unaffected by growth rates in the civilian economy.

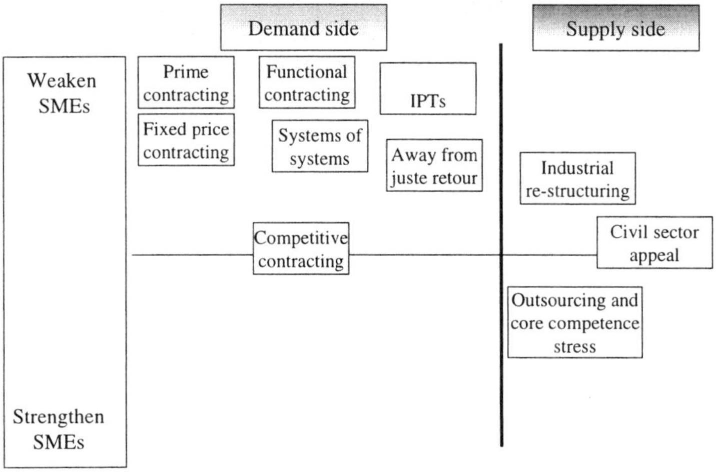

Figure 2: Factors affecting SME's in defence acquisition

As the populated Figure 2 summarises, most acquisition and supply-side changes are making it harder to see that SME's will be more prominent in future defence activities.

3. The way forward

Given that SME's have many supporters with concerns for their future, one option open to governments is to direct prime contractors to direct a specified portion of the value of a contract to SME's. The United States makes heavy use of legal requirements of this nature, enacted by Congress, to push work towards favoured causes such as firms owned by racial minorities. The consequence of such an approach is that it makes defence procurement ever more demanding (US procurement regulations are renowned for their number and complexity). It can also raise the costs and/or performance of equipment if the suppliers chosen are not the most able.

A variation on this theme would be to inform competing prime contractors that a specified use of SME's would be preferred. The maximum weighting that could be given to the SME dimension of bids could then be specified as part of the assessment scheme for dealing with bids. That approach would at least strongly encourage bidders to articulate specifically how SME's fitted into their plans. One constant problem is that bidders which feel they must win a given competition to stay in business are always tempted to offer very low prices that put great pressure on sub-contractors when delivery time arrives after a bid has been successful.

The US, the UK and France spend significant amounts on research and conceptual studies as well as development. Governments that wish to promote SME's in defence can target at least a portion of such funding towards smaller firms rather than passing it all to the major contractors or its own laboratories. The UK, which has a scheme to diversify defence research away from government laboratories and towards universities, is taking some steps down this road.

As important as government practices is the culture that underpins them. In states where much equipment is developed rather than bought off the shelf, acquisition officials must be sensitive to the value of ideas. They must in particular encourage SME's to share their thinking by making clear that a company will not lose out by passing information on a confidential basis to a government. In a situation where a government is focused on the acquisition of valuable but feasible defence capabilities whereas SME's will be primarily concerned with technological innovation, it is plainly desirable that the two dialogues should be in close contact. Each can gain inspiration from the other: one can see that a particular capability will become feasible with the application of a particular technology. The other can look for particular innovations that will facilitate the achievement of a desired capability.

In Britain some dialogue is made possible by the practice of requirements staff exposing some of their thinking in briefings to industry on their future needs. Industrialists hear the briefings, can ask questions in plenary session or can quietly approach requirements staff in the breaks. In principle also, industry can be invited to participate in the Capability Working Groups of the Directors of Equipment Capability in the Equipment Capability Customer area where it can share some of its technological thinking with those responsible for requirements. Procuring authorities can be wary of seeing too close a relationship between the MoD, ECC and industry because it could put one company in a favoured position to win a specific contract. However, the initial route forward here is probably a NAPNOC (No acceptable price, no contract) approach rather than a competition to implement the idea.

For their part, SME's need to work constantly on their relationships with their larger company customers and partners. Partnering arrangements on future projects, including shared risk and funding whenever possible, must become the norm of SME's. Defence sub-contractor status in general and partner status in particular is clearly facilitated if the SME is accredited in quality terms under the ISO-9000 series or some other accepted measure[9]. Appreciation of the qualities of SME's does not negate the basic consideration that, in an era where governments are looking for integrated systems of systems to produce defence advantage, the impact of a single area of technological expertise in an SME is likely to be modest. In defence, SME's are often likely to be distant from the end user, notwithstanding their traditional asset in many areas of being able to stay close to their customers[10]. However, close relationships must not be confined to those further up the supply chain. The British Cogent defence communications company is part-owned by a Canadian civilian telecommunications firm. This firm provides valued access to up-to-date civilian technology leaving Cogent free to focus on its applications and modification for the defence area.

The future for defence-oriented SME's will not be easy and not for nothing have defence businesses across the globe concluded that big is beautiful. The diseconomies of scale pointed out by Alvin Toffler 18 years ago[11] have not proved decisive in shaping the leading defence contractors. Governments wanting to see SME's survive in defence must seek to support them in ways that do not increase the cost of defence equipment, while SMEs themselves must adjust to addressing the changing defence world in which they seek a place.

Endnotes

[1] The author would like to thank his colleague Dr.Derrick Neale for his help in the generation of this paper.

[2] Reference to Matthews et al and the Supply Chain for Warrior. M.Andersson & J.Lilliecreutz, *Supply Chain Strategies and Sub-Tier Structures: Findings from three supply chains in the Swedish defence industry,* Stockholm, DOA (Defence Research Establishment), March 2000.

[3] See, for instance, P.Poutziouris, 'Managing the Small Business' , in S.Crainer (ed), *The Financial Times Handbook of Management,* London, Pitman/Financial Times, 1995, p989-9.

[4] G.Johnson & K.Scholes, Exploring Corporate Strategy, London, Prentuice Hall/Financial Times, fifth edition, 1999, p.29.

[5] See S.Thomke & E von Hippel, 'Customers as Innovators: a new way to create value', Harvard Business Review, April 2002, pp.74—81.

[6] In the 1980's the UK government faced considerable discontent and (almost) legal action from the company that had provided the early thinking for the DROPS concept for delivering ammunition to the battlefield using trucks and a sliding pallet system.

[7] For a basic introduction to the importance of this subject, see B.Rothery, 'Outsourcing' in D.Lock (ed) , *Gower Handbook of Management,* London, Gower, 4th edition, 1998, pp.149-159.

[8] Andersson & Lilliecreutz, op.cit. pp.53-4.

[9] See Rothery, op.cit., p.150.

[10] Poutziouris, op.cit., p.999.

[11] 'the barriers to communication, the lack of manoeuverability, the stifling of innovation, the impersonality, and the loss of motivation that so often accompany expansion beyond a certain point', A.Toffler, The Adaptive Corporation, London, Pan Books, 1985, p.103.

Defence Related SME's
F.D. Carvalho (Ed.)
IOS Press, 2004

From the Soviet Research System to European and International Research Cooperation – Estonian Case

Rene TÕNNISSON
Tartu Science Park, Estonia

Abstract. There are two major changes to which this article owes its existence. The first change is related to the collapse of the Soviet Union and the second to European integration, both in terms of widening and deepening. Those two changes define the overall scope of the current article.

Introduction

These days, Estonia is leaving the past, looking to its future. Paradoxically enough, breaking away from one union and joining another one might be the two most important changes for Estonia during the last fifty years. The following article will describe how Estonia is coping with those changes in the field of research and technology policy.

The last decade of the last century has seen a period of great change, not only for Estonia, but also for Europe and the world in general. Significant political, economic, psychological and other changes have occurred in most countries. Europe is closer than ever before to the peaceful integration of most of its nations into one politico-economic system. At the same time, many are afraid that Europe is developing more slowly than other countries in the world, notably the United States and Japan, because 'losing the competitive edge' is seen as the first step to insecurity, instability, and the loss of living standards. This is why particular emphasis is put on measures which help to keep or increase the pace of development. RTD policy being one of those crucial factors, it benefits from special attention world-wide. A set of measures to secure Europe's competitive position has been implemented, and more will follow according to the decisions of the European Council in Lisbon. The most significant of those are the EU RTD framework programmes which are aiming to better co-ordinate the research activities at EU level, to increase the share of RTD funding, and to promote international scientific co-operation with Eastern and Central European countries as well as with other countries in order to meet the big challenges of the next century: the environment, sustainable development and quality of life.

Estonian research policy should, however, not only be looked at from the perspective of European integration, which is directed mostly towards the future, but also in the context of the past and present situation. Estonia has inherited from the Soviet period both a significant research potential as well as the key problems. In order to adapt the past to current needs and future perspectives, several measures must be taken.

1. Estonia as Part of the Soviet Research System

It is difficult to distinguish the Estonian research system as such from the general research system in the Soviet Union. Estonian research was a well-integrated part of the whole scientific complex of the Soviet Union. Usually this Soviet period is mostly viewed in negative terms. Although there were obvious problems in the Soviet science system, it would be unfair to say that Estonian research in the Soviet Union had only negative aspects; the current scientific potential of Estonia is to some extent a heritage from the Soviet period. However as science is by its nature cumulative and international, the Soviet Union was forced to set up its science system quite similar to the other developed countries and in order to compete with the US. Some scientific fields were quite up to world level. That is particularly true in the fields of natural and exact sciences, which were directly or indirectly related to the Soviet military-industrial complex.

Similarly to other fields of life in the Soviet Union, research was marked by a large degree of centralisation, central planning and hierarchy. Most of the priorities and decisions related to Estonian research, or rather Soviet research conducted in Estonia, were taken centrally in Moscow, without paying much attention to the local needs. Strong centralisation together with central planning of the research made it almost impossible to conduct any national research policy. The only way to get one's voice heard and accepted was through personal contact networks and indirect lobbying in various ministries and research structures in Moscow.

Another feature characterising Soviet research was a rigid institutional separation of research institutes (whether under an Academy of Science or sectorial ministries) from higher education institutions. This separation split the research into two systems, which existed in parallel as research was carried out both in research institutes and universities. However, in general, the research institutes tended to have a higher quality of research than the universities, whereas the researchers working in those institutes were as a rule not involved in teaching.

From the financial point of view, the Soviet research system used a principle of funding research through institutions rather than topics or scientists. This model made it difficult to evaluate the cost-effect structure of the science and to assess the quality of research. Although the percentage of research funding as related to the GDP was not much below the average of OECD countries, the material conditions were bad and salaries of scientists were quite low. Some compensation could have been found in the high prestige and social status of the scientist, which was particularly true in Estonia, where knowledge and learning have always been values of high importance. The research equipment and material conditions for research were poor, with the exception of some institutes which were dealing with priority or strategic fields of study, due to the fact that leading research equipment was often of foreign origin or because it was not sold to the Soviet Union or was considered too expensive.

In 1987, the Estonian research system had a quite significant scientific potential, at least judged by the quantity, whereas the quality of the whole system is much more difficult to judge because of the ambiguity of the available data. At that point of time there were 17 scientific Institutes of the Estonian Academy of Sciences, 18 sectorial institutes under different ministries and 6 universities in Estonia plus about 700 construction offices, laboratories and experimental establishments active in applied research and technological development.

2. Estonian Science Reform

The changes in Estonian research as in most other fields were triggered by political changes in Estonia and the Soviet Union, which took place as an outcome of "Perestroika" and "Glasnost". During the Estonian Science reform, which started in 1990 with the establishing of three new institutions - the Estonian Science Foundation (ETF), the Estonian Informatics Foundation and the Estonian Innovation Foundation, the institutional and the legal basis of the current Estonian Science System were established. The ETF was created in order to implement a new system of research financing through personal research grants, which had never been used before in the Soviet Union. The Estonian Informatics Foundation was set up to develop an Estonian national information system, and the Estonian Innovation Foundation was meant to support the technological development projects. Also, the Estonian Research Council, an advisory body to the government and an umbrella organisation for the three foundations, was created in 1990. Estonian Science reform occurred on a broader scale only after Estonia regained its independence in 1991 and had carried out currency reform in 1992.

The Estonian science reform had to face several issues. On the one hand, the whole structure of research as it had been under the Soviet Union was still there, and it was evident that it was neither possible nor desirable to maintain it, due to very limited resources available and because it was not focused on Estonian realities. On the other hand, it was clear that one cannot close any research institutes or structures overnight but some systematic approach must be developed, which would set up criteria on the basis of which it would be possible to judge the quality of research carried out in various establishments and judge the cost-effect ratio of different research fields. In order to get a picture of the system, the Swedish Natural Research Council was invited to carry out the research evaluation of Estonian natural research in 1991-1992. This evaluation gave a picture of what kind of scientific problems were studied and how well it was done. With the help of this evaluation, it was possible to draw up some conclusions how Estonian research should develop.

The first phase of Estonian research reform ended with the passing of two laws in the Estonian Parliament. Those laws were the Law on Research Organisation in December 1994 and the Law of Universities in January 1995. On the basis of those two laws, the second phase of research reform started with the objective of restructuring the institutional system of Estonian research. The primary goal of the structural reform was to integrate research done at research institutes and universities into one research system where the universities were to be the backbone of the whole system. This phase of the reform had to solve many problems, starting from organisational points, such as on which basis the institutes should be integrated and how much financial and operational autonomy they would keep, and ending with personnel problems related to threats of losing positions as well as personal dislikes of those involved. However, with help of the Law on Research and Development Organisation passed by the Parliament at the beginning of 1997, which replaced the previous Law on Research Organisation, the second phase was formally ended at the end of 1997, when most of the former research institutes had been formally united with universities. The full integration certainly needs several more years. A new institution, the so-called Estonian Science Competence Council, was established under the same law. The main purpose of the organisation is to suggest scope and length of strategic funding for different research fields in Estonian research institutions.

Last year Estonian research strategy – Knowledge Based Estonia - was adopted by Parliament. This document outlines the priority directions for Estonian research policy for the coming years. There is a common understanding that one of the priority directions for Estonia is the integration with the EU RTD framework programmes and with other international research frameworks. A variety of explanations exist why this must be a priority for Estonian research policy. It is argued that there are only two options available: integration with EU RTD programmes, or with those of the Community of Independent States. Estonia can win more from the participation than it can lose. Such integration will create and open up new possibilities for Estonian science, which would help to improve its quality. There is also the possibility of obtaining additional funding for Estonian research.

3. Bridging Estonian and European RTD

Article 76 of the Estonian Association Treaty with the European Union states:

> The Parties shall promote co-operation in research and technological development activities. They shall devote special attention to the following:
> - The exchange of information on each other's science and technology policies;
> - The organisation of joint scientific meetings (seminars and workshops);
> - Joint R&D activities aimed at encouraging scientific progress and the transfer of technology and know-how;
> - Training activities and mobility programmes for researchers and specialists from both sides;
> - Estonia's participation in Community research programmes in accordance with §3. (Estonia's Europe Agreement)

The Estonian Government decided in 1998 that Estonia would be fully associated with the EU Framework Programmes. This decision meant that on the one hand, Estonia has had to pay the participation fee, but on the other hand it has become an equal partner in formulating research priorities and suggesting policy instruments with EU member states. Although Estonia has joined the Fifth Framework Programme on an equal footing with the EU member states, further steps still have to be taken before the effective bridge between the Estonian research system and the EU RTD framework programmes can be constructed.

The process of learning a lesson from the European Union RTD framework for Estonian research policy has begun. Which would be the right way for Estonia and in which way the lesson drawing is currently carried out, is difficult to determine. Different and partially contradictory opinions exist. Some people argue that this lesson drawing is done as a process of synthesis, because the Estonian research system is combining elements from several research systems. It also argued that often some instruments and principles are simply copied from the EU RTD framework without much attention to the local circumstances. It is also thought that the EU should rather be a source of inspiration for Estonian research policy than any direct adaptation. To some extent, all the views might be right and the lesson drawing might occur in several ways, depending on the concrete programmes or policy instruments which are considered.

The general Estonian integration process into the European Union is intensively debated because of two main reasons: to avoid that the past repeats itself and to increase Estonia's welfare and prosperity. Thus, it is the consensus that certain measures in most public policy fields have to be taken in order to let the integration process take place.

In order to achieve the same results as EU member states, a careful analysis must be made of those factors which have been crucial for this development and a prospective evaluation how those factors are related to the Estonian situation would be required in order to try to predict the unpredictable - the future. A lesson must be drawn from the EU RTD system and then applied as a part of research policy. There are several critical aspects which have to be addressed in a proper way in the near future.

For instance:

Estonian RTD institutional structure is to a very large extent missing the industrial RTD, which is one of the main elements of the EU RTD system. As EU RTD framework programmes are strongly focusing on industrial aspects, Estonia might have difficulties to fully benefit from the programmes.

Another critical issue is finance. Estonia is investing into RTD manifoldly less in relative terms than the EU average and more than 10 times less in absolute terms. The prerequisites for successful participation in EU RTD programmes are not only the institutional, human and legal infrastructure but also the material one, consisting of research laboratories and equipment.

The proportions of financing as well as sources of financing basic research, applied research and development activities, do not correspond to the usual proportions of developed countries; they are rather disproportionate and might not allow the achievement of the general socio-economic objectives in Estonia.

Those socio-economic objectives of the research are on the one hand recognised by Estonian scientists and policy makers, but on the other hand the required RTD measures to achieve them are not prioritised enough. The difference in declared and real values must be overcome in order to enable the effective implementation of Estonian research policy.

While elaborating Estonian research policy vis-à-vis European and international contexts, those aspects must be considered and thoroughly analysed and concrete instruments provided to overcome critical issues in order to achieve the prosperity of the Estonian research system and the general welfare of the country.

Defence Related SME's
F.D. Carvalho (Ed.)
IOS Press, 2004

Description of Military Downsizing in Slovakia and Development of New Mechanisms of Cooperation between NATO and Partner SME's

Dr. Stefan VRATNY
BIC Bratislava, Slovak Republic

Abstract. The industrial sector in Slovakia, as a part of the former Czechoslovakia, was characterised to a large extent by heavy industry, a planned economy, orientation to former socialist countries, especially the USSR, and a strong military production. In the "Heavy engineering combine" the main armament company group, more than 80,000 employees were active. After the "velvet revolution", as the communist system was removed, more than 90% of its production was stopped. The impact on the national economy was considerable, the decline of industry production was more than 30%, the unemployment rate went up to 12% within one year and later on even to 20% as most of the supply industry collapsed. The consequences of the military downsizing are not yet solved; following an expert estimation, it is not realistic to remove them before the end of the decade. The decay of industrial production also caused a decline of the national research and development. In addition, the social consequences have been considerable.

NATO SME's could subcontract the partner countries' SME's both in military and in civilian production. That would improve the cost structure of NATO SME's and correspondingly help the partner SME's to increase their output.

The SME's could transfer their technologies both from NATO SME's to the partner ones and vice versa. There are various technology transfer possibilities, from licence agreements, through technical assistance and subcontracting manufacturing to joint venture creations.

NATO SME's could invest either directly or through financial institutions into the partner SME's. The investments could be carried out also in the form of technology equipment, providing of managerial and/or marketing skills, etc.

1. Impact of military downsizing in Slovakia

The industrial sector in Slovakia, as a part of the former Czechoslovakia, was characterised to a large extent by heavy industry, a planned economy, orientation to former socialist countries, especially the USSR, and a strong military production. In the "Heavy engineering combine" the main armament company group, more than 80,000 employees were active. After the "velvet revolution", as the communist system was removed, more than 90% of its production was stopped. The impact on the national economy was considerable; the decline of industry production was more than 30%, the unemployment rate went up to 12% within one year and later on even to 20% as most of the supply industry collapsed. The consequences of the military downsizing are not yet solved; following an expert estimation, it is not realistic to remove them before the end of the decade. The decay of industrial production also caused a decline of the national research and development base. In addition, the social consequences have been considerable.

1.1 The general situation

The problem of military downsizing in Slovakia was however more complex. There were three factors that affected the national economy in a very negative way.

After 1990 the ancient socialist system broke down. The so-called planned economy was removed, but a new market economy has not been introduced. The most public, especially State orders were stopped immediately after the "velvet revolution". This has brought instability not just to the manufacturing sector but also to the Research and Development (R&D) base in Slovakia. All companies at that time were State owned, and the management lacked experience of a market economy. The privatisation process was not done in the right way, the State companies were privatised through a "voucher privatisation" which did not solve either the ownership or the capital input needed for companies. The next significant problem was and partially still is, an insufficient knowledge of international markets, a lack of marketing skills and radical changes of export markets.

Slovakia in frame of the former Czechoslovakia exported almost 40% of its production to the former Soviet Union and less than 11% to Western European markets. That has changed dramatically; exports to the former Soviet Union dropped to less than 1%. Fortunately, exports to Western markets, especially to the European Union, went up to 60%.

1.2 Military production downsizing in Slovakia

The end of the Cold War was a logical consequence of the breakdown of the previous socialist system. This clear positive impact affected the economies and enterprises in many countries, especially however in Slovakia.

The share of the armament production during the socialist system was about 25% of the whole GDP. The production was partially financed by the Stated-owned budget and partially by exports especially to Warsaw Pact countries, but also to some Asian and African countries. The largest concern in Slovakia involved in armament production was ZTS "Heavy engineering combine". ZTS was the main armament company group; more than 80,000 employees were directly employed in the ZTS companies

And about 300,000 further employees were involved in delivery and service activities. This figure of 380,000 workers represented more then 15% of the entire working population of the Slovak Republic. After the "velvet revolution". as the communist system was removed, more than 90% of its production was stopped immediately due to a wrong political decision. The impact on the national economy was considerable; the decline of industry production was more than 30%, and the unemployment rate went up to 12% within one year. Consequently most supply and service companies also collapsed and the unemployment rate rose to a dangerous 20%.

The consequences of the military downsizing can be identified in the following sector:

a) Decline of the GDP and of industrial production: The GDP declined between 1990 and 1993 to 34%. The currency was devalued; the decline of the industry sector was even larger. The inflation rate rose considerably in that period. It lasted 11 years until the GDP level from 1989 was reached. The State budget balance generated deficits.

b) Decline of the R&D base: The national R&D base dropped significantly. Just one-third of the previous R&D base remained as State orders and industrial orders, driven mostly by armament production, shrank. The reform of the R&D base has not been carried out yet and the country suffers heavily under brain drain as the number of young educated people leaving the country increases.

c) Social impact: The most significant impact of military production downsizing in Slovakia is the social one. The unemployment rate rose extremely fast from 2% to 12% within one year in 1992 and later on even to 20%. The regions where military manufacturing facilities were based suffer under a depression. The last factory involved in military production collapsed early this year. Due to the following extremely high unemployment whole families have suffered, especially if both couples worked there. This social situation is also affecting children. Poverty is typical for those regions. Nation wide more than 11% of the population is suffering from poverty.

d) Political impact: The socio–economic situation also has an impact on the political situation. As a consequence of the collapse of former military production and the following high unemployment, populist parties gained political power in 1992 and remained until 1998 in leading positions.

1.3 How to solve the problem of downsizing military production

The consequences of military downsizing are not yet solved; following an expert estimation, it is not realistic to remove them before the end of the decade. That means the impact of military downsizing will last at least two decades. However, there are tendencies leading to improvement of the situation described. The most important is:

The conversion of military production

A part of the conversion of military production has already been done. The conversion can be characterised as follows:

a) Implementation of new industrial sectors: Military production in Slovakia almost disappeared but it was successfully replaced by the automotive industry. Although the share of the automotive industry in the national GDP is still smaller than the former military production, it is about 15% compared to the former 25%; this share is increasing continuously and could reach 25% by the end of this decade or even sooner. The output of the whole mechanical industry is already about 80% compared to the period of military production and will probably reach the original level within the next five years, particularly due to the increase of the automotive industry.

b) Creation of technology-based SME's in regions depressed by military downsizing: There were three main regions involved in military production. Two of them are already partially converted, the third one collapsed recently and recovery has not yet begun. In the first two regions, new production has been introduced, e.g., hydraulic components, production of gearboxes and motors, car components, etc.

c) Attraction of investors to the depressed regions: Special infrastructure measures such as industrial parks, development of investment stimuli for foreign investors and the expectations of a relatively near membership of NATO and of the European Union are preconditions for attraction of direct foreign investments. It is necessary to bring these investments to regions depressed by military downsizing.

2. Proposal of new mechanisms of cooperation between NATO and Partner SME's

2. 1. Manufacturing cooperation

NATO SME's could subcontract the partner countries' SME's, both in military and in civilian production. That would improve the cost structure of NATO SME's and on the other hand help partner SME's to increase their output. There are more possibilities for mutual co-operation. A relatively easy way is to subcontract a part of the manufacturing activities to partner SME's. In Slovakia there is special expertise where manufacturing could be subcontracted. The most important sectors are:

- Automotive sector, especially the supply clusters
- IT sectors, especially Software development and programming
- Wood processing sector and furniture manufacturing
- Energy generation, especially renewable energies

2. 2. Technology transfers between NATO and partner SME's

SME's could transfer their technologies both from NATO SME's to partner ones and vice versa. There are various technology transfer possibilities, from licence agreements, through technical assistance and subcontracting manufacturing to joint venture creation.

There is one special system working in Europe, in 30 European countries. It would be worthwhile to think over the possibility of extending this also to the USA. The European system is called Innovation Relay Centres – IRCs.

The IRC Network is a network of centres hosted by well-known regional organizations providing specialized business support services for innovation and technical cooperation in Europe, e.g., in the EU, in associated countries and in candidate countries.

A network of 68 consortia contains almost 250 partner organizations, thus ensuring easy geographic accessibility to 30 European countries in Member and Associated States of the EU. Nearly 1,000 experts are involved in activities supporting the development of trans-national technological co-operation especially among small and medium-sized enterprises.

The scope of technology transfer activities among SME's can be demonstrated by the example of IRC Slovakia.

The most important activities and services are the following:

a) Technology Transfer Support Activities: The Innovation Relay Center Slovakia supports clients from the Slovak Republic in the trans-national technology co-operation with the EU and associated countries.

Within the framework of the support of national and trans-national technology transfer IRC Slovakia is providing a broad variety of activities for its clients, from contacts exchange to consultancy in technology transfer agreement preparation. Networking is probably the most important activity – exchange of contacts within the frame of European networks – partner organisations in EU and associated countries - goal-directed international meetings of companies – Technology Transfer Days in particular listed areas - insertion of technology offers and requests into the databases (European-wide exchange system of technology offers and requests – BBS, database of tool producers, database of automotive industry suppliers, database of companies evaluating the biomass) - complex technology diagnostics of the company – technology audit and intensive support in carrying out the resulting technology transfer in the case of a possible match of offer and request.

b) Targeted International Meetings of Companies: One of the main instruments of implementation of the technology transfers methodology is a trans-national brokerage event

- Technology Transfer Days (TTD). The objective of these undertakings is to create a common platform for co-operation between companies willing to innovate their technology process. The purpose of TTD is to support the transfer of technologies between organisations and companies in Slovakia, EU countries and associated countries.

TTD's are organised in all major technology sectors in Europe and the IRC. Slovakia specializes in the automotive industry, information technologies, plastic and wood-processing industries and the agro-food sector.

These events are targeted to small and medium-sized enterprises as well as to the larger companies that are offering or looking for technologies, tools and solutions for their products, services, development and manufacturing processes or commercial activities in the respective fields. Usually it is a good opportunity for participants from Slovakia to establish new business contacts with potential partners from the European Union and foreign companies have the possibility to meet a significant number of Slovak companies active in the relevant field of industry. The system for technology transfer and technology exchange could be possibly extended to other NATO and partner SME's, as well as the USA.

c) The BBS system: another tool, which is used for support of trans-national technology transfer is the electronic system developed by the Central Unit in Luxembourg. This system called the Business Bulletin System is aimed at exchanging technology offers and requests among companies in the European Union and associated countries. Nowadays the system contains more than 4,000 technology offers and approximately 1,400 technology requests. Registered requests and offers are divided into ten basic technological groups and technology codes are allocated to each offer and request. These are then spread throughout the IRC's among companies from the relevant sectors in 30 countries in Europe.

Slovak companies are allowed to register in the system through the IRC Slovakia website; they choose their technological area and then automatically receive the relevant technology offers and requests from the system by e-mails. It is also possible to register their own technology offer or request through forms which are available on this website. Operators of IRC Slovakia then place these offers or requests into the BBS system and these are spread among companies in 30 European countries.

d) Services: IRC Slovakia enhances the competitiveness of companies by filling technology gaps or by marketing innovative technologies throughout Europe and helps to find the right partner for technological co-operation among the companies from relevant areas.

Other services provided by IRC Slovakia are audits and technology watch. Their aim is to stimulate the capacity to adopt new technologies (technology audits, IPR, novelty scans, technology watch, organise SME visits, etc).

IRC Slovakia also assists in the negotiation of a trans-national technology partnership agreement, facilitates access to EC and EUREKA research results and to the European Innovation initiatives, e.g., in innovation financing, venture capital, start-up or in Intellectual Property Rights areas.

2.3 Investment mechanisms

NATO SME's could invest either directly or through financial institutions into partner SME's. The investments could be carried out also in the form of technology equipment, providing of managerial and/or marketing skills, etc.

The transfer of capital is the most complex co-operation form between NATO and partner SME's as it can include subcontracting, technology transfer or an acquisition of whole companies. There are some examples mostly from the large company sector. One of these is the acquisition of the Eastern Slovak Steel Works by US Steel (USS). Two years after the acquisition the Slovak steel work is the most profitable in the whole USS group world-wide. There are some other examples such as IKEA or Volkswagen, where the investment in Slovakia is one of the most profitable ones in the whole concern. The question is, however, how the SME's could carry out these types of investment, as they often do not have relatively large capital resources for acquisition activities.

One possibility could be the establishment of special investment funds, similar to the Venture Capital Funds, which would invest in technologies and manufacturing equipment in their own countries, e.g., in the EU or in the USA and then transfer them, in a cooperation with local NATO SME's, to partner SME's. NATO SME's could also provide management and marketing know-how and could acquire shares in partner SME's. After partner SME's were developed, the Funds could exit either through the sale of their shares to NATO (or a Partner SME or to both) or via a stock exchange. That construction would bring together the capital needed for NATO SME's, the technologies for partner SME's and the yields from these investments could be achieved through a combination of NATO SME's marketing know-how and partner SME's manufacturing know-how. These types of funds could be developed both in the European Union and in the USA where public investment facilities like the EIB, the EBRD or the IFC could also be involved.

The Third Power

Han WILLEMS
NIID, The Netherlands

Abstract. Politics, Primes and SME are the key players in military procurement.

Politics uses every opportunity to praise SME for their cost saving potential and to point at the necessity for Governments and Institutions like the EU to assure optimum use of SME assets. However the procurement of major weapon systems only involves Primes. Whether they involve SME is no issue for the Procurer in Europe.

Europe does not have "The State of Small Business". Why not??

Make or Buy decisions become more relevant in the growing consolidation of the European Armaments industry.

Taxpayers are entitled to see that all possible means are used in the most cost-effective way, therefore the RFQ procedure must include SME. All windows of opportunities must be used, including a European SME database.

Also Primes may gain from using such database and create a WIN situation for themselves. The same applies for the Procurement officer, also he can WIN by making absolutely certain that SME is included to the maximum

SME is WINning by being included in the procurement process, ending up with a three times WIN situation.

Can NATO promote the inclusion of SME in SMC?

Can NATO assist in the creation of ETSA, the European Transnational Small Business Act?

The third power of WIN by using THE THIRD POWER of SME is something which is not to be ignored.

Introduction

When someone gets the opportunity to speak up in the interest of small and medium-sized companies in Europe, he has to realize that he speaks for many, each with an interest of its own. That does not make life easy, let me try however to deliver something useful for most of them.

As advisor to NIID, the defence industries branch organization in The Netherlands, my daily business concentrates on SME and SMC (SMC stands for Small and Medium-sized Country). I am also Chairman of the SME committee in EDIG, called SME-DIG. In both capacities I have three personal motives:

1. Make sure that every policymaker and business manager is well aware of the strength ***and limitations*** of SME, especially of SME in SMC;
2. Assure the optimum use of SME assets in European Nations;
3. Eliminate bureaucratic "support tools" and put generic ones in place.

You might wonder about the title of this address. There is more than one way to use this mathematical expression. I start with one and I will conclude with others.

The first power is **POLITICS**. This is true in normal life and even more in the military world. The second power, sometimes greater than the first one, is the **PRIME**

CONTRACTOR. Major R&D in long lasting programmes is the true basis for their strength in creating large complicated systems. The third power is SME.

Politicians and other governmental officials use almost every opportunity to praise SME, using many nice words. Let me give you a few examples of these words:

- "Small business is the lifeblood of our economy"
- "Does the EU think small first?"
- "If Governments do not think small first, they may never think small at all"

In these words you see a breakthrough, it is the recognition that Governments and institutions like the European Union have a key role to play in ensuring the optimum use of SME assets.

During this seminar we question whether NATO is also in that position. Before starting any discussion on SME, we have to realize that SME is best explained by Small Management Enterprise.

Real SME has little personnel for R&D, to explore new markets, to answer the many questions of Governments; in short to do the thousands of things to be done in business life. There is one CEO, who also is the sales agent, the troubleshooter in the production etc, etc. Asking this person to attend meetings equals the question to him of giving up some business. Asking this person to fill in a lengthy questionnaire equals asking him to skip at least one telephone call with a potential customer.

The European Commission acknowledges this to be one of the problems in dealing with SME:

Small business entrepreneurs may have better things to do than sit on advisory committees

And this puts us in a dangerous position, because as usual we address the topic of SME without SME being present.

In my opening statement I gave you my legitimization to speak up for SME, I wonder what yours will be?

Let me share with you some thoughts on three issues

1. Government
   a. As enabler
   b. As procurer
2. Prime contractor
   a. As integrator
   b. As producer
3. SMC
   a. Nation without real Prime contractors
   b. SME to rely on participation, offset and above all on export.

1. Governments struggle with SME.

Many governments realize that SME is one of the important engines in the national economy. Creating and maintaining the right conditions for enterprises, large and small, is generally accepted as a task for Governments.

This justifies assistance in several areas, like R&D, export, finding partners, etc. This justifies also support by grants, subsidies and guarantees.

The biggest struggle however is HOW to reach SME.

Mostly Governments try to do this by putting a specialized agency in place. Millions of Euros are spent for organizations to assist SME to obtain the assistance they really need.

It is bizarre that so much effort and money is put into a system that intends to help and in reality creates a real administrative burden on SME.

Are there alternatives?

It is certainly not easy; however I think there are.

Using tax systems for support reduces the burden AND provides some control on applying the rules on a proper way.

Another one is the creation of a user-friendly display of opportunities.

A European industrial database is a first step. The European Commission is supporting the creation of such. Brian Morris, who will speak to us tomorrow, is best placed to elaborate on this tool.

Even more important than the initiative to put something in place, is the control on the effective use. For example: What is the meaning of a database if its use is limited to the organization tasked with the maintenance of it?

The United States of America found a way, a very long time ago.

It was called: **"The State of Small Business"**; a report of the President of the United States of America transmitted to Congress, as early as 1987.

This early date is a clear indication of the time lost in Europe. Allow me to come back to this when I address SMC. For now my question is: "Why should Europe not need similar legislation?

This brings me to the role of procurer.

Since most of us are convinced that SME provides real value for money, it should be the task of every procurement officer to assure that SME is participating to the maximum. This will save expenditure.

The procurer should seek maximum involvement of SME by direct procurement using open competition.

He also must force the prime to open fair competition between in-house capabilities and outsourcing, in case the order has to be given to a real prime contractor for technical reasons.

In both cases the procurement officer must play an active role:

- Controlling that SME is included in the RFQ procedure
- Checking whether consultation of the Euro-SME database revealed the names of the companies to be included in these processes.

One of the most important decisions with great impact on SME is the MAKE or BUY decision.

The United Kingdom found a way to deal with this dilemma. So-called Codes of Practice were agreed between Government and Industry. Codes deal with many issues, one of them is the choice between MAKE and BUY. It is rather peculiar that this excellent initiative is not yet implemented in the UK itself, nor in any other nation or international organization.

Let me conclude this part with five recommendations:

1. Use wherever possible generic instruments to assist SME
2. Promote the European SME database
3. Prescribe the consultation of this database in the procurement
4. Open discussion to adopt NATO codes of practice
5. Prepare for a NATO wide legislation system dealing with the use of SME assets in NATO nations.

2. Large companies

So-called primes also face some problems with SME.

How to find the best partners in SME, how to include them in the logistics of the production system, how to assure quality, how to deal with classified information and many more HOWs.

In the consolidation processes the large companies get more in-house capabilities, potentially enabling an increase in efficiency. This potential gain is overshadowed by the overhead system that controls such large entities. As long as the prime can include this overhead in the quotation, he will do so and use his own in-house capabilities. Doing so, competition is nearly eliminated and sometimes misused by asking quotations to enable the in-house price to be compared.

The decision for Make or Buy must be and can be made more transparent. The UK Codes of Practice foresee this.

Many Primes complain about the offset.

Offsets have to be provided to other nations. Apart from the phenomenon of offset, which is a subject of its own, there is a real danger for SME in the nation asking for offsets to be excluded.

The prime will seek a larger company in the "offset nation". This so-called lead company is supposed to make arrangements for all companies to be involved in the offset programme. When this company decides to keep most of the offset within the company itself, competition is excluded again and SME remains empty-handed.

Primes seem to be well advised to seek SME partners in many nations, partners with state-of-the-art capabilities, partners being included in the logistic chain of the system integrator. The better the Prime takes care of this, the sooner offset is not an issue any longer.

3. SMC and its complications

Many nations have no national prime contractor; some have none at all while others have none in specific sectors like aerospace, maritime or land-based systems. SME in such nations are depending on export, on participation and on offsets. Those SME have to break through the nationalism, through the security of supply argumentation, they have to deal with national export rules, which are sometimes more stringent than in the nation of the prime contractor. They have to cross borders for most of their business. It is most important that the capabilities of SME in SMC are accessible in the European SME database. It is even important that the governments of the SMC nations demand the consultation of that database for each procurement in a foreign nation.

Finally it must be assured that SME in the SMC is included in the production, when possible also in the development of the weapon system. For this reason Europe should consider implementing a law, similar to the State of Small Business in the USA.

I t could be named the:

European Transnational Small business Act - ETSA

It should prescribe that SME in the buying nation must be included in the process. This means that the prime has to put real effort in finding at least two SMEs, capable to meet the demands on competitive terms.

For SMEs in a closed market such as the defence business, it is a real and serious problem to find out when any MOD is ready to start new procurement or when a major player has received an order to which the SME can contribute. National associations have a key role to gather the information and to pass it on ***only*** to relevant SME. This is done already on a national basis, what we really need is to do this on a European basis. Today I dare to ask for a NATO-wide basis; however Europe should first get its act together before we can effectively cross the ocean.

NATO may consider promoting the early involvement of industry associations in SMC nations.

Last but not least, an international procurement office, similar to OCCAR, should replace the inefficient national procurement systems as soon as possible.

When such agencies perform according to the economic laws they must send an RFQ to SME.

Even in the closed environment of procurement agencies, it remains valid to state that involving SME potentially brings a **WIN** situation for the procurement officer.

4. Conclusions

Ladies and gentlemen, I could go on for hours. I just touched upon a few issues and I did that in a brief format. The discussions we will have during these days will certainly cover more and in more detail.

I am very pleased to hand over the floor to a friend for many years, a real professor, who will share with you the view of a prime.

He will show you how appropriate the selection of SME by primes leads to a **WIN** situation for the prime.

Finally this session will end by a friend from the periphery of NATO. He will address SME in SMC; a typical example where a WIN situation for all parties is within reach.

In doing all of this we end up with a **WIN x WIN x WIN**, which equals **WIN**^{3}

Let me conclude with the same mathematical expression once more: I do hope that the wordplay on the screen will help you to remember the great power of SME.

The Survival Power of SME's in an SMC: The Turkish Case

Faruk Aga YARMAN
Chairman, Defence Industry Sector, TOBB
(Turkish Union of Industry, Trade, Bourses Chambers)
Vice President, SaSaD (Turkish Defence Industries Association)

Abstract. In this presentation an attempt shall be made to assess how SME's are positioned first within their national economy, from the perspective of a small defence industrialist trying to survive in an SMC.

Following a discussion on SME's in SMC's from the national point of view, in general the picture shall be further extended towards regional, European and the NATO context with an emphasis on defence industries. For the purpose of selecting Turkey as a case for study, a review of the role of SME's in the national economic life shall be performed, along with some typical government incentives available.

It will be explained that, despite their limited size, allocated funds are far from saturation, probably due to the mismatch of the lethargic and cumbersome nature of the bureaucratic processes vis-à-vis the efficiency and dynamism inherent in SME's spirit. International Organizations may concentrate on persistent efforts to promote policies and strategies to support and develop SME's, but global priorities are not always set with such consciousness. We know that, when Primes are well-off, so are their favoured subcontractors, mainly SME's. But then, what happens at a crisis time? Who gets the priorities on international and national grounds?

Therefore, to really promote SME's, in addition to the traditional governmental/international, support/assistance/incentives programmes some radically different approach is needed, especially when dealing within the frame of defence projects. Not surprisingly every nation is trying to protect its defence industry base. But obviously these nations which house the major global defence contractors are luckier than others. The rapidly widening technological and economic gap between the leading nations of NATO remains a constant source of irritation within the Alliance and consequently a potential niche for politically destabilizing forces.

A healthy, hence long-standing partnership is only possible among equitable friends. In a global single-poled equilibrium, peace will only last if it is fed by a full symbiosis that embraces a common, uniform distribution of wealth, leading to harmonized, integrated defence industries. If the main defence contractors originate mostly from large nations, it is about time for the SME's of SMC's to become their favoured suppliers. When achieved, a fairer distribution in workload shall surely improve the underlying national, regional, continental and transatlantic harmony in NATO.

Introduction

From a national perspective, SME's are the major building blocks of the economy and social life and when organized, a determining factor of our political sphere. But their voice is only heard if they face the survival problem collectively. Obviously under normal operating conditions their preoccupation to survive (due to their size, efficiency and instinctively isolated focus) is pushing them to struggle not only with volatile local economic conditions, but also to reinforce themselves against competitors at the other end of the world; whom they actually do not know, but smell the existence of, thanks to the globalization winds! Like a biological cell, in an ecology made of technological, economic and social elements, SMC's are engaged in a lively rhythm. If redundant they first compete, then crowd out each other; otherwise they coexist in an harmonious environment. They are small and mostly efficient, hence they ought to survive by definition. If they find the favourable biosphere they grow to a certain size, then multiply by simple division. Sometimes they grow or merge to something else. In case of crises however, they first stop their multiplication, then endure with a minimal life, then start dying or become absorbed, to leave the ground for the stronger. The secret of their strength lies in their micro-cosmos made of multifunctional employees who are ambitious, productive and entrepreneurial by nature.

Traditionally, until the Industrial Revolution, they had higher visibility not only in economic life, but also in shaping the social structures of the states, like individuals in societies. Centuries later, now that we are surfing in the information age, not surprisingly special attention is needed to understand their microbiology and interrelation to the super structure that we call Main Contractors, be it national, regional or international!

They are so self absorbed in their own world, so busy surviving that often they do not feel that actually everyone is for them.

1. SME's Role in the Turkish Economy

SME's are positioned first within the national economy as the backbone of industry. Just a quick look at the figures shows that 99.2 % (in numbers) of more than 200,000 enterprises in Turkey are SME's employing less than 150 workers. This corresponds to say 56% of the total workforce in industry who produce merely a quarter of all the national industrial added value. Although this ratio is quite low by OECD's standards, still more than 20,000 SME's are providing nearly 40% of the national export. As a developing large nation, unfortunately and traditionally SME's focus on production with cheap labour, and use only low or medium-level technologies (e.g. textiles, food, chemical and metal industries, etc.), the only exception being the defence and IT side. This general picture can easily be extended towards European and other NATO countries. Looking at European countries statistics one can place Turkey in the mosaic, comparing the number of enterprises to those of say Denmark, Ireland, and Austria. For average employment, compare Turkey's 8 to that of France, UK and even Belgium and Denmark.

Everyone is favouring SME's, as is national government. The Turkish political and bureaucratic elite have spent more than a decade "in good faith" to really promote and support SME's. In the mid-80's, "KOSGEB" (Small and Medium Industry Development Organization) was established by the Ministry of Industry and Trade. Soon after, a countrywide organization followed to serve and support provincial industrial zones, in order to provide financial and technical assistance, support R&D, IT and e-commerce activities, regional development, export promotion and competitivity enhancement and fostering of entrepreneurship. In order to initiate and coordinate these activities, the Headquarters of KOSGEB was housed in the biggest national defence Main Contractor MKEK (a major player in the field of weapons and ammunition). The aim was to upgrade the role of medium-level industries in meeting the social and economic needs of the country vis-à-vis the globalizing competitive world. It was reckoned that such an attempt should enhance their competitive capacity to facilitate their integration in industry and conform to economic development.

Following KOSGEB's foundation, in order to realize these objectives, Services Centres were created throughout the country. They specialized in: consultancy and quality improvement, sector specific development (machinery, electronics, computers, materials, chemical, medical, automation, etc.), market research, common facility workshops, training, technology development, investment guidance and information.

Later, the government also initiated a series of actions for financial support and assistance (using resources from both national and international investment institutions) in SME's investment (e.g., investment credits, tax and duty exemptions, energy support, credit guarantees, land provisions) support to improve global competitiveness and R&D (on project basis, tax rebates, risk capital, institutional supports).

Not to downplay all these sincere efforts, but frankly, from an industrialist's perspective, I believe that it is still quite difficult to claim that the results are satisfactory. Am I spoiled? Maybe flattered but not spoiled! Then why complain?

First of all, in all these support/assistance/promotion programmes, fire and water come together. That is to say the dynamism of the wild SME and the lethargy of the tamed "state". At the end of the day is not the taxpayer the ultimate support? Yes. These are SME's and their workers! Secondly, no matter how cynical the idea is, my ex-boss used to say: "help those people whom you wish to paralyze!" Thirdly, what an SME ultimately wants is the freedom to compete or to die! Therefore all we need to provide to the SME is work in competitive conditions (instead of national, regional or continental discrimination). These ideas of course need some calibration when wishing to align different companies of different regions at different maturity levels, due to different historical backgrounds.

Reality lies in natural solutions, and if required, only limited control should be allowed for transition periods. Hence perhaps some kind of code of conduct may provide a sound solution; a certain balance is kept, but never artificially supports inefficiencies being sustained forever. This situation reminds me of the picture of the loving young relatives, insisting on artificial life support (in an expensive intensive care room of hospital!) for the senior member of the family who wisely prefers an end to his pain and reserve limited resources for the healthier younger ones.

2. My Post-modern Mood and the Impact of the Information Age

In the last two decades, in parallel with the end of the Soviet Union, the Information Age has picked up. Unlike the agricultural or industrial ages, nations have to go beyond their traditional resources (i.e., land, workforce and capital). They now have to promote their intellectual existence, cultural transformation and their development in harmony with all these innovations. The atmosphere should enhance the vital role of SME's by giving them a unique status not only in their local environment, but also as working members of our planetary family. Globalization has been forcing Primes and their sub-contractors for the development of new concepts, sharing of innovations and information and pushing to a new order. The big nations who took part in the creation of this new age now have to deal also with subsequent problems at home and abroad. At home for example following the disaster of September 11, we now know that terrorism is no longer a regional issue in problematic countries; it always was and is and shall be a global matter for the years to come. Terrorism is fed by both spiritual and material misery!

Leaving aside the emerging power of the Asian Tigers, in the last two decades, the ex- "Western World", the US and Western Europe have been living with the dilemma of mutual support and competition in trade. The in-continent merger fever finally has reached cross-Atlantic mergers and take-overs, leaving us with a problem: how do we complement transnational Primes indiscriminately with the SME's of our world. For the new order is not problem-free. Increased integration levels in the global economy are also for sure increasing the imbalance between the rich and the poor at individual, national, regional, continental and intercontinental levels.

The asymmetry between the US and the EU has long been felt; it is now imposing itself between the large nations of the EU and the SMC's or large developing countries. One of the major problems that globalization and subsequent liberalization is bringing to our agenda is that the continuous and merciless competition between companies is also inspiring them to strategies where destruction of some competitor is preferred to integration. Again an artificial singular point of the new order. Considering the asymmetries of nations in the distribution of wealth, no wonder some countries fearing to lose may prefer a limited opening-up or even worse, tightening the isolation of their national economies.

Opening of national economies to globalization therefore requires careful planning of the process in order to avoid losing control of national priorities in national industrial complexes and hence, national sovereignty.

3. Turkish Defence Industry Base and SME's:

I start this section by adding more political flavour: on the one hand, Turkey is at the cross-roads of three continents and has a tradition of optimizing its geo-politics with its geo-economy (even in most volatiles stages of a long-standing history). A loyal ally in NATO and a continuing candidate for EU membeship, Turkey is the most Eastern country; at least this is the way the modern Republic perceives itself. Turkish national interests obviously also cover the Middle East, the Balkans, the Caucasus, Central Asia, the Black Sea and the Mediterranean Region. Turkey has a population of 60 million, of which 5% already live in EU countries. The Turkish economy ranks within the first 20 large economies of the world and has a competitive structure with a young population who are qualified and entrepreneurial in spirit.

Now turning to the Turkish Defence Industry Base and SME's; in Turkey as in many other SMC's, the local defense industrial complex actually consists of SME's. National defence leaders (in general "majority of shares being under the state control") are in the field of weapons and ammunition, missiles, electronics, C4I, automotive and armoured vehicles, aeronautics and shipbuilding. The Turkish working model relies on "National Main Contractors".

Moreover, the Turkish defence industry depends heavily on the transfer of technology from her allies. Hence the local players are either subsidiaries/participants of global players (eventually trying to capture the local market), or national companies teamed up with global players on a defence project basis. As a result generally in Turkey, the SME's focus rather on the civilian market and their contribution to Turkish defence contracts are unacceptably weak. Not only is the local value added in defence projects low, but the Turkish defense industries' contribution to the civilian marketplace is also low, so reminding one of the old Soviet model.

This situation has two dangerous implications; firstly there is an unbearable load on the Foreign Trade Deficit due to defence procurement programmes; secondly, and even worse, a disassociation of the national industrial complex in general from the national defence industry. This situation could be even more dramatic because of the possibility of an alienation of the national economy from the larger NATO nations and thus from the global economy.

Although the working model relies on National Main Contractors, the Turkish defence industry depends heavily on transfer of technology from allies. The players are either subsidiaries/participants of global players, or they have to team up with them on a project basis. As a result, in general the SME's focus in Turkey is mainly civilian and their contribution to the Turkish defence contractors is extremely weak. This in turn has two dangerous implications; the Foreign Trade Deficit in defence matters is far from bearable and even worse, the Turkish general industrial complex is disassociated from the local defence industry (reminding us of the ex-Soviet model) in the first place and totally isolated from global defence contractors mainly consisting of the larger countries of NATO.

4. Closing Remarks

The collapse of the Soviet Union towards the end of the last century has shaken the global defence market, forcing the supply side to restructure.

Market forces initiated similar trends on both sides of the Atlantic, pressuring large companies to merge, thus leading to large defence contractors becoming smaller in number and lots of specialized SME's.

At the turn of the Millennium this radical restructuring has shaken the global defence industry, leaving a picture where 80% of the world defence equipment market is supplied by less than a hundred Main Contractors, mostly from the US, the UK, France, Germany, Italy, i.e., the larger countries of NATO. The procurement agencies of the leading nations in NATO pursue value for money while optimizing life-cycle cost, performance delivery and quality in meeting their requirements. Therefore they prefer single-stop shopping with the Primes who in turn do the same with their familiar SME's. For the small contractors, the best way to break into the defence market then became the option of acting as the subcontractor to prime defence contractors. On the other hand, despite all modern trends, nations keep their protectionist procurement restrictions.

Needless to say, the status quo obviously is in favour of SME's in the neighbourhood of global primes. The widening gap between the market share of SME's in conjunction with their geographical location is alienating SME's of SMC's from globalization trends. Artificially promoting incentives, support/assistance programmes are far from satisfactory. A code of conduct in the Alliance may help to solve the problems discussed in AC/313 in NATO.

The decreasing number of global defence contractors also generated a sizeable number of SME's focusing on their narrow activities for works to be subcontracted. These SME's have a vital role in the overall competitiveness through their contribution to the technological base of their respective SMC's. In a world at peace they cannot be left aside.

SME's would then become expert companies who provide components or a niche product, manufacturing parts or offering specialized services, leaving the ground for stronger integration in the world, along with the overall responsibility of quality, configuration and programme management. And all of this in a much fairer distribution of work.

Table 1: Economic indicators for SME's in Turkey

Number of Workers	Enterprises	%	Employment	%	Added Value	%
1-49	199.338	97.9	722.763	42.8	865.8	12.9

50-150	2.578	1.3	221.226	13.1	758.0	11.3
150- +	1.630	0.8	743.303	44.1	075.7	75.8
Total	**203.546**	**100**	**1.687.298**	**100**	**6 699.5**	**100**

Table 2: National Eximbanks Credit

	Number of Enterprises	Amount (bn$)
Eximbank	3319	3.300
Amount reserved for SME	2124	1.188
Percentage	%64	%36

Table 3: SME's in the EU

Total Number of Enterprises:	16.350.000 (99,78% SME)
Total Employment:	101.350.000 (%71.48 SME Employment)
Average Employment:	6
SME Work Force Productivity:	45.000 Euro/Worker
Added Value/Workforce Cost:	59%

	Scale	Number of Enterprises	Number of Workers	Average Employment
Micro	0-9	15.210.000	31.450.000	2
Little	10-19	605.000	8.250.000	13
	20-49	370.000	12.250.000	33
Middle	50-99	70.000	4.950.000	71
	100-249	60.000	10.400.000	173
	250-499	15.000	5.100.000	340
Big	500 (+)	20.000	28.900.000	1445
Total		16.350.000	101.350.000	

	Number of Enterprises (x1.000)	Average Employment	SME Weight	Work Force Productivity
Belgium	490	6	SME	108
Denmark	170	9	SME	100
France	1.980	7	SME	90
Germany	2.290	9	SME	102
Greece	715	3	Micro	94
Ireland	180	8	SME	91
Italy	3.920	4	Micro	90
Luxembourg	15	10	SME	94
Holland	420	10	SME	99
Portugal	600	5	SME	88
Spain	2.460	4	Micro	92
UK	2.630	8	SME	83
Austria	180	12	SME	79
Finland	150	12	SME	98
Sweden	150	13	SME	95

GLOBAL PLAYERS IN DEFENCE *
First 30 Companies with Defence Revenues exceeding 1 Billion USD

Country	***Number of*** Prime Contractors	*2000* Defence Revenue **$Billion**	***2000*** Total Revenue **$Billion**	Percent of **Revenue From Defence**
U.S.A	**16**	**84.5**	**351**	**24**
EUROPE	**10**	**33.0**	**78**	**42**
JAPAN	**2**	**4.7**	**33.5**	**14**
RUSSIA (Export)	**1**	**3.1**	**3.1**	**100**
ISRAEL	**1**	**1.3**	**2.1**	**62**
TOPLAM	**30**	**126.6**	**467.7**	**25**

* Defense News Top 100-2000, July 2001

OCCAR COUNTRIES AND TURKEY

Country	Revenue **Billion €**	**Employment**	Revenue Per Capita **k €**
UK	**24**	**175**	**137**

FRANCE	**11**	**68**	**166**
GERMANY	**4**	**35**	**125**
ITALY	**2**	**14**	**159**
TURKEY	**0.820**	**11**	**75**

Chaillot Papers

Country	**Equipment Sales Mio USD**	**R & D Mio USD**	**%**
U.S.A.	47	35	74
UK	**8.2**	**3.9**	**48**
FRANCE	**5.2**	**3.1**	**60**
GERMANY	**3.7**	**1.2**	**32**
ITALY	**1.9**	**0.5**	**26**
TURKEY	**0.36**	**0.012**	**3!**

Million USD, Chaillot Papers, 2000

Looking at the Defense News, Top 100-2000 list, we can make the following remarks at the turn of the Millennium :

- Within the 100 companies, the defence revenues rank from $18 bn to less then 10mil.
- Altogether the top-100 (excluding China) add up to $151 bn of defence revenue.
- 83% of the $151bn comes from the first 30 companies who make more than $1bn/year in defence.
- 12% of the $151bn comes from non-NATO countries, with the contribution of Russia (4%), Japan (3%), Israel (2%). Other players with less than a 1% contribution are Switzerland, Sweden, Singapore, Canada, Australia, Turkey, India, etc.
- Major defence contractors in NATO who make more than $1bn/year in defence - US (56%), Europe (22%) can be seen in the Table below. * Defense News Top 100-2000, July 2001.

Issues and Opportunities Facing SME's in Partner Nations Today

Prof . William ZADORSKY
Ukrainian Ecological Academy of Sciences, Ukrainian State University of Chemical Engineering, Pridneprovie Cleaner Production Center
Dnepropetrovsk, Ukraine

Abstract. The main opportunities for SME's in Partners nations today are:

1. To increase the significance of SME's in the solving of Sustainable Development Tasks (Economy, Ecology and Social Problems in Integrity)
2. Processes of SME Restructuring, Privatization and Military Conversion in countries with transitional economies, taking account of market conditions.
3. Technological Business Incubation (TBI) as the tool and method of SME development. Here the most usable directions (approaches) are Technological Business Incubator ("Warmhouse") Technology as the tool and method of SME development.
4. Modern Information and Internet Technologies for SME development and cooperation.
5. Education and Training of military-oriented SME staff in Sustainable Development, Cleaner Production, Pollution Prevention, Energy Saving, Industrial Symbiosis and Life-Cycle Assessment directions.
6. Advanced Technologies for SME development.

Introduction

The main opportunities for SME's in Partners nations today are identified in the following six paragraphs of this paper.

1. To increase the significance of SME's in the solving of Sustainable Development Tasks (Economy, Ecology and Social Problems in Integrity).

The sustainable development of a country is possible only by the simultaneous balanced solutions of economic, ecological and social problems, as was marked in the Agenda for the 21st century accepted at the Summit of leading countries of the world in Rio de Janeiro. SME's act here in two ways; on the one hand, they have to have balance solving and the solving of economic, ecological and social problems simultaneously. On the other hand, they should execute concrete ecological, economical and socially-oriented projects.

World experience testifies that the main tendencies in maintenance of sustainable industrial development of an industrial region are: development of low- and non-waste technological processes and equipment and salvaging industrial and household waste. The ecological concept of manufacture lies in the basis of the system approach .

Thirty years ago American scientist Barry Commoner formulated four ecological laws: (1) everything is interconnected; (2) everything should have its place; (3) everything has a cost; (4) nature knows better than us. The ecological interrelation of all kinds of natural resources requires a system analysis of their condition, a complex forecast of use and definition of allowable volumes of their use.

It would be possible to consider the interaction between nature and man on the basis of a complex systematic approach founded in the belief that the technical equipment is only a part of the overall system. Hence the tendency to harmonize the relationship of nature with technical equipment, where the operation of industrial complexes is tied not only to the technogenic activity of man and the use of technological objects, but also when the state of the natural environment becomes evident. The ideal solution to the problem would be the creation of nature-technical systems allowing the achievement of high technical indices at favourable ecological conditions.

In so far as it is frequently impossible to reduce the level of negative influence of manufacturing on the environment without a change in technological processes, nature-saving activity should be directed at improvement, or on the creation of predominantly new technological processes, directed not only at deciding on helpful problems, but also on protection of an environment.

It is important to have the measure of a quantitative evaluation of the degree of stability of one or another project. It is the main Sustainable Development problem for SME's – simultaneity and synchronism of problem solving and the necessity of the concrete realization of every project.

Thus, it is necessary to have a system of quantitative evaluation of the ecological relation of industrial objects and the degree of an ecosystem's destruction in the acceptance of justified decisions, directed by the ecological impact of industrial and agricultural manufacture and rehabilitation of territories polluted by toxic substances. An important problem is the integration of processes of acceptance of decisions in the field of an environment and development, improvement of a given system and its analytical methods. Complex measures in evaluation of the consequences of decisions in economic, social and ecological spheres are necessary. It is necessary not only at a level of separate projects, but also at a level of a policy and programmes. The analysis should include evaluation of costs, profits and risks.

From the point of view of the system analysis of the problem of interaction of a person with the environment in accordance with development of an industry, feedback begins to play a more and more significant role, i.e., the influence of an environment on the development of manufacture. In the opinion of many experts in the field of industrial ecology it is necessary to aim not only at completing clearing or waste salvaging, but also to admit to their occurrence, and achieve non-waste or, at least, cleaner technology. In any technology it is necessary to evaluate the degree of its ecological danger, i.e., on the quantity and quality of waste formation. Parameters of quantity and quality of pollution of emitted gases, of waste water and of solid waste are the most objective criteria of technological imperfection.

For example, Slavin M. proposes [1] to use economic effect E for the realization of technical arrangements directed to environmental protection, in accordance with the quantitative evaluation mentioned above. It may be calculated by the formula:

$$E = \int_{i=1}^{i=10} E_i \qquad (1)$$

Where:

E_1- value of supplementary products or energy resulting from utilization and use of waste and lateral side-products;

E_2- economy of means expended for transportation production and end result of waste in places of centralized destruction, transformation or utilization;

E_3- economy of means expended for remediation of chemical and ecological ground and reservoir structures or for liquidation of negative consequences owing to the impossibility of their rational use;

E_4- liquidation of negative economic consequences of ruin or toxification of industrially-used animals, wild and cultured plants, the result of which makes their use inadmissible or inefficient;

E_5- economy of means expended for medical treatment, diagnosis, preservation and nursing of invalids through poisoning or toxification, caused by unfavourable influences on the environment;

E_6- reduction of means, which the State expends for payments for temporary disability due to disease owing to unfavourable influences on the environment;

E_7- liquidation of negative economic consequences of cost of products unobtained owing to loss of working ability of persons in the zone of the production influence on the environment;

E_8- economy of means expended for buildings' external appearance and restoration, gardens, parks and other places for workers' rest and recuperation and restoration of architectural buildings and memorials in the care of the State;

E_9- economy of means expended for planned and unplanned environmental condition control realized by sanitary epidemic stations or other sanitary and hygiene services;

E_{10}- liquidation of negative consequences through product cost unobtained due to temporary halts in production by the violation of recommended sanitary and hygiene standards.

Ecological influence criterion (e.i.c.) allows us to evaluate the degree of production technology perfection from the viewpoint of interaction with the environment; this is proposed by V.Anikeev with co-authors [2]. They suppose, "the most representative and useful ecological criterion is the ecological influence criterion k (e.i.c.)", which defines by formula:

$$k = W_t / W_r = W_t / (W_t + W_c) \qquad (2)$$

Where:

W_t- theoretical influence necessary for production;

W_r- real influence;

W_c- influence determined by concrete production.

The maximum significance of e.i.c. equalling 1, is determined by condition Wt = Wr; that is when real influence corresponds to the necessary theoretical level which is determined by laws of substance and energy. The lower e.i.c. requires more exact technical solutions, from the view-point of the influence on the environment. And when $k \rightarrow 0$, the production does not take into account the absolute demands of guarding the environment.

It is also proposed [3]*indexes of ecological efficiency of technological processes as follows:

- Ecological measure (L) - quantity of harmful action on the environment (P), divided by the quantity of useful production or services one gets with the help of this process (Q):

$$L = P/Q \tag{3}$$

-Resource capacity of the process (N) - debit of energy, water, air, ground and other natural resources (R) to the useful production or services that one gets with the help of this process:

$$N = R/Q \tag{4}$$

- Ecological index of the object (E) defined by the formula:

$$E = (Q - P)/ R \tag{5}$$

Where (Q - P) is the useful effect.

All sizes of Q, P, R are obtained in a natural significance; it seems to the authors that the ecological index of the object (E) describes the degree of the object in close relation to nature. When $E = 0$, the potential of nature destroys without any useful effect, and when $E = 1$, there are no non-utilization remainders of matter or energy.

Thus, it is necessary to have a system of quantitative evaluation of ecological relations of industrial objects and the degree of destruction of ecosystems for acceptance in justified decisions, directed by the ecological impact of industrial and agricultural manufacture and rehabilitation of territories polluted by toxic substances.

As for Wotte J. [4] a precise way of evaluating technological processes in order to develop and to introduce Low and Non-Waste Technology (LNWP), theoretically would be by cost-benefit analysis. In this connection not only technological and economic influences should be taken into consideration but also ecological and social aspects. The question of monetary evaluation of environmental impacts and similar effects, however, has not yet been solved at the international level. Therefore other ways must be sought for evaluating technological processes.

New methods were tested [4] and completed especially for industrialization purposes in developing countries. The methods developed may be used:

- To compare several technological processes (at least two) for the production of a defined product,
- To evaluate the processes from the standpoint of LNTW,
- To make a clear choice-of which process variant should be preferred because of its technological, economic, ecological and social characteristics, depending on national economic conditions as well as regional or local prerequisites.

The developed method is based on selected characteristics, which are grouped in a minimum programme, characterizing the technological process and, if necessary, the aim of the main product from the standpoint of low-and non-waste technology (LNWT). The working steps of the selection represent a combination of evaluating as well as weighting the given characteristics.

For evaluating the characteristics against the background of a world level analysis, the estimates for each characteristic are:

-The Zero-value, K_o (accepted worst value) and

-The Best Value, K_b (best possible value).

The evaluation is calculated by the equation:

$$PW = \frac{K_o - K}{K_o - K_b}, \quad (6)$$

Where:

K - Real-Value of the given characteristic. This delivers values for all characteristics for all technological variants under discussion within a given industrialization project.

For weighting the characteristics, a group of experts is formed consisting of technologists, economists, ecologists and specialists in territorial planning. All the experts should be skilled specialists in their profession or their fields of activity respectively, with long-time experience, the private interests of which are not connected in any way with the interests of the enterprise ordering the selection procedure.

After complete information from the experts' group on advantages and disadvantages of the technological variants under discussion as well as the conditions within the region and the whole national economy, each expert receives a questionnaire which contains the characteristics grouped for a paired comparison.

Each expert separately marks each pair of characteristics; which of them from his point of view in the given case should receive a higher importance. On the basis of the experts' decisions made by this paired comparison a personal preference table is compiled for each expert.

The judgements shown in all the experts' decisions are tested by well-known methods of mathematical statistics. First of all, for optimization of the natural environment, an ecological and hydrochemical evaluation of the objects of optimization, using normative limiting parameters of environmental quality and methods of ecological optimization of surroundings, are required. There are both forecasting and regulation of optimal processes, valuation of efficiency of use of mineral resources, control of the use of nature, regional economy and ecological monitoring with its technical programme and information maintenance. And, lastly, the control and management of quality of the natural environment on the basis of the science of the biosphere and its evolution are required.

The ecological interrelation of all kinds of natural resources requires a system analysis of their condition, complex forecast of use and definition of allowable volumes of use of natural resources. A complex quantitative evaluation of a level of sustainability of the development of a system expediently uses the integrated parameters of sustainability, which may be defined as a multiplying coefficient including the main technical characterizations of the technological process connected with the resulting measure of production purity such as: flexibility of production; intensifying of the technique; efficiency of the technology, etc., and also parameters, that reflect, accordingly, social and economic efficiency of transformation, determined by expertise. Integrated parameter S (parameter of sustainability), is determined as the product:

$$S= aE \cdot bR \cdot cK, \quad (7)$$

Where:

a, b, c - "weights" contribution of appropriate parameters,

R, K - parameters, reflecting, accordingly, social and economic efficiency of transformation, determined by expertise,

E- parameter of a system's ecology, which is expedient to define as proceeding from the following reasons.

In our opinion, quantitatively, the degree of the production improvement as regards the influence on the environment, may be estimated as a multiplying ecological coefficient:

$$J = \prod_{i=1}^{k} J_i , \quad (8)$$

Where:

J_i are the main characteristics of the technological process connected with the resulting measure of production purity such as:

J_1 - flexibility of the production;

J_2 - intensiveness of the technique;

J_3 - efficiency of the technology, etc.

Any of these measures may be defined similarly to the efficiency of any technological object as:

$$Ji = \frac{P_{1i} - P_{2i}}{P_{maxi} - P_{2i}}, \quad (9)$$

Where P_{1i}, P_{2i}, P_{maxi} -are respectively the final, initial and maximum measures.

The value P_i may be both a common index characterizing quantitatively one or another property of the system that connects with corresponding measures (for example, dust collecting efficiency, extraction ratio, etc.), and a complex integral index, which counts several characteristics of the object simultaneously.

In the latter case one is supposed to use either additive indices:

$$Pi = \int_{j=1}^{n} K_j P_{ij} , \quad (10)$$

Where K_j is the significance of the j-th indice (estimated expertly and changing within the limits of O to 1).

It is necessary that the multiplying ecological coefficient J for comparison of the different technological installations contains the identical typesetting J_i.

Connecting these databases, approaches, concepts, etc., in the joint system, is possible only by theories and methods of technical creation and with the use of laws of the development of technical systems, of the use of psychological peculiarities of creative activities and modern methods of creative technical activity in solving problems of ecology (such as a method of attempts and errors, brainstorming, morphological analysis, synechtics, theory and algorithm of inventory tasks' solution, decision-making in risk conditions, elements of the theory of usefulness, means of support of acceptance of decision creation of information systems in analysis and modelling of ecological situations; methods of the analysis of conditions; forecast of development; modelling of the scripts of decisions on waste problems; a means of expert evaluation of proposals in the decisions on ecological problems; integrated base of knowledge of waste problems).

The defensive concept of protection of nature and population in the Ukraine is today replaced by a new concept, the basis of which is not only transformation, ecology, sustainable development of agriculture and industrial complexes, but by adaptation and rehabilitation of the population. On the basis of SME's, it is possible to realize all complex problems of the integrating of economic, biological and human systems in collaboration between engine-, info- and mathematical modelling and eco-communities.

There are a number of excellent examples in North America and other countries where SME participation in ensuring sustainable development of a country shows that it is a cost-effective strategy anywhere for encouraging environmental improvement with simultaneous SME prosperity:

1. Recently Sunoco, Inc., announced that 500 companies endorsed the Coalition for Environmentally Responsible Economies (CERES) Principles and reinforced its long-standing commitment to safe, reliable and environmentally sound operations. Sunoco believes that excellent performance in health, environment and safety goes hand-in-hand with operations excellence and financial strength. All employees have ownership in improved results in each of these areas.

CERES, a unique partnership between some of America's largest institutional investors and environmental groups is a worldwide leader in standardized corporate environmental reporting and has pioneered an innovative approach toward encouraging greater corporate responsibility on environmental issues. Over the last decade, a number of other major corporations have joined with Sunoco through their endorsement of the CERES Principles. Sunoco, Inc., with headquarters in Philadelphia, is a leading manufacturer and marketer of petroleum and petrochemical products. With 730,000 barrels per day of refining capacity, over 4,150 retail outlets selling gasoline and convenience items, interests in over 10,000 miles of crude oil and refined product pipelines and 35 product terminals, Sunoco is one of the largest independent refiner-marketers in the United States. Sunoco is a growing force in petrochemicals with over nine billion pounds of annual production capacity, largely chemical intermediates used in the manufacture of fibres, plastics, film and resins. Utilizing a unique, patented technology, Sunoco also manufactures two million tons annually of high-quality blast furnace coke for use in the steel industry.

2. "Green Star" is a type of programme of local government that promotes Cleaner Production and environmental improvements by business and other organizations. A Green-Star type programme run by a local government or other entity promotes Cleaner Production (CP) to local business by:

a) Getting a voluntary commitment by the business to participate,
b) Encouraging progressive levels of CP, recognizing accomplishments with Green Star awards (or similar),
c) Providing technical assistance,
d) Public recognition when business achieves a level of progress.

For example, the programmes in King County, Washington and in Boulder, Colorado each include over 300 companies, and each is staffed with only one full-time local government worker.

A lot of local governments, especially in developing countries, operate such programmes. The study of their experience will evaluate the cost-effectiveness of this particular promotion model.

3. Recently the Ukrainian On-line Community for NGO's http://www.iatp.donetsk.ua/~NGO-net also announced a new subject "Technologies of Sustainable Development". At present the concept of sustainable development is being widely discussed in Ukraine. This concept means integration of economy, ecology and social stabilization. What initiatives can be put in place to achieve economic development? What initiatives can be used to achieve ecological development? What initiatives can be taken to achieve social development?

2. Processes of SME Restructuring, Privatization, Military Conversion in countries with transitional economies taking account of market conditions.

During Restructuring, Privatization and Military Conversion for SME's, there are inconsistencies between requests for maintenance of sustainable development and the problem of deriving maximum profits. In countries with a transitional economy, there is unfortunately a slacking of State and municipal management and no marketing mechanisms for managing SME activity. Recently a new programme on Market Mechanisms and Incentives for Environmental Management (URL: http://es.epa.gov/ncer/rfa/02marketmech.html) as part of the U.S. Environmental Protection Agency (EPA) and its Science to Achieve Results (STAR) were announced. They are seeking applications for research leading to improved theoretical and/or empirical analyses of the feasibility and effectiveness of market mechanisms and economic incentives (MM&I) as substitutes for, or complements to, traditional environmental management programmes. The terms 'market mechanisms' and 'incentives' refer to approaches that rely on economic incentives, market forces, or financial mechanisms to encourage regulated entities to reduce emissions, discharges and waste generation, or generally improve environmental performance. The EPA is interested in supporting research on practical applications of MM&I approaches related to its mission, i.e., addressing environmental quality and human health – The Applicable Category(s): Grant/Fellowship Announcements.

3. Technological Business Incubation (TBI) as the tool and method of SME development.

The start period is the most difficult for SME development (they need investments to enter into the information market, to acquire the technologies and equipment, to build, etc.). Bank investments often are too burdensome for SME's. Here the most usable directions (approaches) are the Technological Business Incubator ("Warmhouse") Technology as the tool and method of SME development.

We launched such a TBI and now our specialists and scientists can represent new activity directions and new technologies for a new time.

TBI envelops a wide range of technical domains. Except for "classic" consulting, TBI gives the following supplementary benefits to a client:

True Professionalism: the clients will be attended by better experts in technology;

Global Reach: will be used as a true global examination by information experts of the Internet;

High services: receive speed thanks to Internet possibilities;

Low Expenditure: because of extraordinarily low cost of fittings.

Today SME's TBI is drawing together all stages of elaborative marketing:

- Patent and legal services
- Business-planning, professional preparation and interpretation of publicity material
- Project presentation on the Internet and the mass media on international conferences, exhibitions and fairs
- Search of potential partners and conclusion of partner agreements
- Search of potential investors in Ukraine and abroad and detection of investment agreements
- Grants search, participation in tenders, competitions, conforming to project subjects, professional claims preparation
- TBI also provides consulting services for organizations, companies and professionals on commercial beginnings. Services can be from answers on information inquiries to marketing, information and technical decision search and search also of partners and investors, design of concrete objects or projects
- Ecological and technological examination and audits of industrial enterprises (chemical, metallurgical and machine-building profiles)
- Marketing, organization and publicity campaigns on production, including briefings for journalists, television and broadcasting, specialist publication in newspapers and magazines
- Decision on problems of Cleaner Production, Waste Management, Pollution Prevention and Energy Saving on the basis of System Approach and Sustainable Development
- Post-graduate retraining of specialists in problems of Cleaner Production, Waste Management, Pollution Prevention and Energy Saving
- Accumulation and transfer of information on decisions taken through experience of technological and ecological problems using resource saving and low level waste technologies
- Elaboration and realization of investment programmes on actual Cleaner Production themes

By means of experienced consultants of TBI, SME clients are able:

- To have access to more than 800 million Web pages with vast information on business, production, trade, education, problems of society and policy, science and techniques, medicine and health, culture and art, problems of family, home and leisure, etc.
- To open free of charge a mailbox of electronic mail and to receive and to send correspondence
- To open free of charge and to formalize an electronic Web page
- To become acquainted with multimedia novelties and free of charge to be accorded the latest files of computer guaranteeing, programmes, games, etc.
- To avail themselves of the possibilities of contemporary computers and Internet for foreign language learning, for dealing with computer programmes, for working with the Internet, etc.
- To acquaint themselves with musical novelties and the world (sights of treasure-houses, better museums, historic places, light wonders, etc.),
- With use of Internet, to find possibilities of grant and other subsidies for education abroad
- To buy goods in Internet shops at any point in the world
- To pick up jobs or professions abroad and to formalize and to send by electronic mail the necessary documents

- If necessary to avail themselves of the service "electronic matchmaker"
- Without supplementary payment to have a talk on the phone with relatives and friends in other NIS countries; if by computer there is proper equipment and the installation of necessary programmes
- To open minds and simply have a rest with computer games (music, acquaintance with literature, visit to any world library, etc.)

In the Internet-Business-Centre (IBC), SME businessmen, by means of TBI consultants, will be able:

- To find partners and search out companies for collaboration
- To find goods and service producers
- To know the prices of goods and services
- To obtain information on financial work indexes of potential partners abroad
- To prepare Internet information about their own organization, suggestions for collaboration, projects, etc. with a calculation of international requirements
- To search out potential investors and investments, to acquaint themselves with the very latest information on grants, competitions, tenders and other investments
- To obtain and to complete application forms for all sorts of investments
- To participate in business, scientific and technical Internet conferences on professional questions
- To become acquainted with legislation in any country in the world on questions related to their professional interests

IBC plans to deliver:

- Firms and organizations, goods and service presentations
- Business negotiations at a contemporary international level (simultaneous translation, contemporary communication funds, other services)
- Briefings for the mass media
- Internet conferences
- Meetings with leading guests, scientists, specialists, interesting people, seminars, conferences, sales exhibitions of contemporary engineering and programme software.

4. Modern Information and Internet Technologies for SME development and cooperation.

As SME's have a rather small staff, they can solve the problems of development and realize industrial activity only because of cooperation and only by using the possibilities of modern information technologies and, first of all, the Internet with strong navigation methods and Global Search Machine systems.

5. Education and Training of military-oriented SME staff in Sustainable Development, Cleaner Production, Pollution Prevention, Energy Saving, Industrial Symbiosis, Life-Cycle Assessment directions.

Up to now, military-oriented SME staff have not been oriented to Sustainable Development, Cleaner Production, Pollution Prevention, Energy Saving, Industrial Symbiosis and Life-Cycle Assessment directions. It is necessary to provide special education and training for those who are re-orienting. Two of our main training programmes for SME's are listed below:

5.1 Business on basis of information technologies

- Basis of system analysis and theory of decision making (4 hours)
- Design of management on the basis of information technologies (4 hours)
- Using modern information technologies and their possibilities in the Internet for organization and development of production (2 hours)
- Search engines in the Internet – navigation strategy - searching of patent, scientific, commercial and other information, partners and investors (4 hours)
- Development Strategy of technological business-incubators and particularities of a functioning virtual technological business-incubator (2 hours)
- International requirements for the contents and registration of websites and pages, basis of web design (4 hours).
- Preparation of business offers and plans, brand technologies (4 hours)
- Work on international investment and innovative markets. Legal regulation of intellectual property protection. Investment markets (2 hours)
- On-line training on navigation in the Internet; work with search engine servers. (4 hours)
- Performing exhaustive work on one of the following: preparation of business offers/plans, advertising avenues and presentations, project websites, searching for partners and investors in concrete projects (16 hours)

5.2 Information technologies and specialist education of VIPs, administrators and businessmen

- Basis of computer and Internet literacy (8 hours)
- Hygiene in using a computer; methods of eliminating bad influences of computer technology (technicians) on people (2 hours)
- Using modern information technologies and possibilities such as the Internet in specialist education (information-search systems to network the Internet – strategy of record retrieval -patent, scientific, commercial, etc., partners and investors, using methods of educating, remote and on-line education overseas, searching subsidies, scholarships, etc., Internet education, foreign languages, management and marketing (4 hours).
- Modern business using information technologies. Searching partners and investors. Development of a strategy of technological business-incubators and particularities of the virtual technological business-incubator. International requirements for web site contents and registration, Web design bases (without using language HTML), preparation of business offers and plans, brand technologies, work on international investment and innovative markets (8 hours).
- On-line training on navigation in the Internet (8 hours). Total - 30 hours

6. Advanced Technologies for SME development.

One of the most advanced technologies for use in military-oriented SME's is nanotechnology. Recently the US EPA started a special science programme on this topic: "Environmental Futures Research in Nanoscale Science, Engineering and Technology"

(Title: Environmental Futures Research in Nanoscale Science, Engineering and Technology) - URL: http://es.epa.gov/ncer/rfa/02nanotech.html. The U.S. Environmental Protection Agency (EPA), Office of Research and Development (ORD), National Center for Environmental Research (NCER), as part of its Science to Achieve Results (STAR) programme, is seeking grant applications for research on nanoscale science, engineering and technology, collectively referred to as nanotechnology. The EPA is particularly interested in proposals in Nanoscale Science, Engineering and Technology that address environmental issues in four topic areas of potential application: Environmentally Benign Manufacturing and Processing; Remediation and Treatment; Sensors; and Environmental Implications of Nanotechnology. In the fourth area, 'Environmental Implications of Nanotechnology' proposals are sought for smaller grants - The Applicable Category(s): Grant/Fellowship Announcements).

Some of the author's advanced technologies for SME development will be offered such as Commercialised Technologies Virtual Market (Some commercialized technical and scientific proposals of more than 300 of my patents and inventions are listed here. Mutually beneficial cooperation in these projects is possible) :

For agricultural SME's:

- Electrosol spraying for agricultural applications
- New Approaches to Spraying Pesticides
- Commercialized environmentally friendly technologies:
- Extension of Ecological Concept of Cleaner Production
- Adaptation and rehabilitation programme of inhabitants of ecologically-intense zones
- Method and equipment for washing fine dispersion solid phase products
- Ecological Equipment and Environmental Protection
- Elaboration of the ecological concept of cleaner production
- Electrosol Apparatus and Methods
- Natural and Wastewater Treatment Technology and Instrumentation
- Improving water quality through low-cost upgrading of a water treatment facility accompanied by abandonment of chlorination
- Pollution Control Equipment and Environmental Protection
- Pulsed Voltage Converter for Electrical Filter Power Supply
- The programme and experience of education retraining in theory and practice of cleaner production
- Treatment of Natural and Waste Water
- Algorithm of production purity increasing on the basis of a system approach with RSP's using:

For Manufacturing SME's:

- Catalyst impregnation
- Flexible Production Systems
- Electrosol painting
- Impregnation of electrodes and other carbon/graphite articles
- Impregnation of textile fibres and fibre reinforcement of resin-matrix composites
- Solid-liquid extraction for pharmaceutics, food processing and pulp industries
- Flexible Manufacturing for Chemical Processing
- Electrosol Equipment and Methods
- Dioxane production (Increasing of production purity when using RSP's)
- Applications of parallel reaction-separation processes (RSP) as cleaner production (CP) technologies in chemical engineering

For SME's producing Equipment and Materials:

- Acoustic Equipment for Intensification of Mass-Transfer Processes
- New Impregnation Thermocondensed Technology of Production
- Technology for deep purification of liquids
- Acoustic Devices for Increasing Rates of Mass Transfer Processes
- Electrosol Apparatus and Methods
- Flexible Manufacturing Systems for Chemical Processing
- Pulsed Voltage Converter for Electrical Filter Power Supply
- Valve Contact devices for mass-transfer equipment
- State-of-the-Art Approaches to Aerobic Fermentation Engineering

References

[1] Slavin M. Social-hygienic factors of new technique // "The questions of economics".- Moscow: 1979.-N5.-p.58-66

[2] Anikeev W.A., Kopp I.Z., Skalkin F.W. Technological aspects surrounding defense.- Leningrad.: Gidrometeoizdat, 1982.-254 p.

[3] Environmental protection. The models of social-economical prognosis.- Moscow: Economica, 1982.-224 p.

[4] Wotte Joris. Selection of Technological Processes from the Standpoint of LOW AND NON-WASTE TECHNOLOGY for Industrialization Projects. Journal of EIA Vol.3, No.2 November 1994, p.9-14.

Defense Related SME's: Analysis and Description of Current Conditions
F.D. Carvalho (Ed.)
IOS Press, 2004

New and Existing SME's Need to Cope with Emerging Civilian Markets

Krzysztof ZASIADLY
Business and Innovation Centre,
Poznan, Poland

Abstract. In Poland, Russia and Ukraine, alongside civilian employees losing work in the process of military base closure, jobs are needed also for the officers, non-commissioned officers and their family members, as the conversion is accompanied by a reduction in the Armed Forces as a whole.

Due to the author's experience with social adaptation programmes in Poland, Russia and Ukraine, the problems are similar in the three countries mentioned.

Retired military personnel face many problems: psychological, medical, social, financial, adaptation to new environment, etc. The Army and the Government are not able to entirely provide all their needs.

There are different programmes developed for military basis closure, military retired men and their families, which differ from country to country and depend on the local environment. Some of them are focused on entrepreneurship promotion. Military men have many advantages useful in business: strategy thinking, team building, responsibility, logistics, etc.

However, it is not easy to develop business today. The bad condition of economies leaves small enterprises without much support and bankruptcy results.

Vocational training and psychological aid offered by social adaptation programmes for retired military personnel in Eastern European countries is not enough. In Ukraine and Russia it is supplemented by aid programmes focused on families of military personnel. However, there is a need for more active programmes for entrepreneurship promotion.

Even military base closure generates new opportunity. This process needs different activities which can be conducted by companies managed by retired military personnel.

Social Adaptation programmes in Poland, Ukraine and Russia could be more effective if they were more closely linked to military property ownership restructuring. In Poland both programmes are managed by the Ministry of Defence. That may be an advantage in the future.

Facilities could be offered to develop business incubators and industrial zones. Revenues collected from selling military property should be used to support development of military small business, including loans, technical assistance, training, etc.

Eastern European countries should collaborate and develop networks to exchange experience and know-how regarding social adaptation and small business development.

Governments and Armies in Eastern Europe should care more about retired and released military personnel than they do at present. They should learn from Western countries how to do it.

New NATO members should adapt NATO standards regarding the policy towards retired and released military personnel as well as military base reconversion.

Related to restructuring of the Army is also the problem of military industry, which has been cut in Eastern countries. It means hundreds of thousands unemployed and idle facilities.

Introduction: Military base closure

There are two systems of conversion of military bases in Eastern European countries. The first deals with conversion of large military bases, actually mono-functional cities. The second one refers to small military towns, including remote ones. As a rule there are small settlements located near small regular towns. These military towns are most often transferred to the local self-government institutions, but not enough funds for their conversion are allocated by the State. And local authority has no money for that either. And it is a set of problems: ecology, housing for people, use of human resources, absence of any business, etc.

Military base closure can have a catastrophic effect on a community, both economically and emotionally, wiping out thousands of jobs and uprooting friends and neighbours. Therefore Governments and local administration strive to protect people and facilities. After 1990 hundreds of thousands of military men were released in Poland, Russia and Ukraine. Members of their families multiplied this number. Hundreds of military towns were disarmed. Reorganization also affected dozens of other facilities and people.

If handled wisely, the loss can become an opportunity to diversify a local economy. In some cases, Polish cities have emerged from the process not only unscathed, but also stronger and more vibrant. Its administrators considered a range of options.

1. Case study – City of Borne Sulinowo, Poland

Poland passed the first phase of base closure when the Red Army left for Russia in the early nineties.

Borne Sulinowo in Northern Poland, not wasting any time, began taking over buildings and land almost immediately to keep them functioning and allowing for a smoother transition. They were literally moving in as troops were moving out.

The community of the region welcomed the new administrators; there was solid support from the regional and governmental administration, residents and the area's representatives in Parliament.

Additionally, the administration launched a local public relations campaign in an effort to attract investors and potential inhabitants. The administration gave presentations on the relocation project. There was not a target group that they had not brought in to tour the complex. The Polish government kicked in millions of Polish zloties due to Reconversion of Military Base and Ecology Protection Programmes.

They wanted this to be successful so they could point to it and say, you can close bases and it can be good for the community. It was hard to close the Army Base with its 30,000 inhabitants, soldiers and their families generating jobs for the surrounding civilian community.

Finally the complex became a capital of regional administration, and includes a hospital, nursing home, kindergarten, centre for people from polluted Polish regions, hotels and conference facilities, recreational facilities and a town centre with restaurants, bookstore and additional amenities.

The city is designed to house companies, which can take advantage of the natural resources. So far, more than 260 companies have set up shop in the city and another 200 in the neighbouring area. The city maintains a Business Incubator with a Training and Consulting Centre as well as an Entrepreneurship Loan Fund on the site created to recruit new businesses. Since 1977 the alliance has brought more than 70 companies to Borne

Sulinowo. And unemployment in the area is now low despite economic development problems in the country.

It is possible that Borne Sulinowo had a lucky break going through the closures during the economic boom times of the 1990's. The economy was good; the Borne Sulinowo area was experiencing massive growth.

Rebounding may not be as easy if the Polish economy continues to slow and it will not be easy to replace lost jobs if the unemployment rate reaches dangerous levels.

Stories such as those of Borne Sulinowo are welcome news at the Polish Ministry of Defence, where officials are eager for evidence that a community's survival after the loss of a military installation is a matter of success.

A second phase is now underway. Polish Army bases are under a closure process. About 65,000 officers, soldiers and their families have to be socially adapted, employed or self-employed and transferred to other regions.

2. Military entrepreneurship and social adaptation

In Poland, Russia and Ukraine, in addition to civilian employees losing work in the process of military base closure, jobs are needed also for the officers, non-commissioned officers and their family members, as the conversion is accompanied by a reduction in the Armed Forces as a whole.

Due to the author's experience with Social Adaptation programmes in Poland, Russia and Ukraine, retired military men in the three countries mentioned face many problems: psychological, medical, social, financial, adaptation to a new environment, housing problems, and consequently an inexpediency in the resettlement of families from military towns to places where no free houses are available. The Army and the Government are not able to entirely protect them and their families. In Poland the position of retired officers and their families is a little bit better than in Ukraine and Russia due to higher GNP per capita and salaries.

Poland, Ukraine and Russia develop different programmes for military base closure, retired military personnel and their families. They include also NGO's. In Russia the "Partner" non-commercial Foundation of Social Adaptation for Discharged or Retired Military Servicemen, their Family Members and Citizens of Other Categories, works within the framework of the programme "Social Adaptation of Discharged or Retired Servicemen". In Ukraine the International Foundation for Social Adaptation (IFSA) started when the Retraining for the Military Programme terminated in 1998. Both programmes are confirmed by the Open Society Institute of New York.

Depending on the local environment, programmes differ from country to country. For many years the Polish Military Centre for Professional Activization did not deal directly with entrepreneurship promotion. However it was financing business courses upon the individual request of officers who wanted to start their own business. They also could get additional seed money to start their business. The Centre now intends to start a couple of Business Incubators using idle military facilities.

The "Partner" Foundation in Russia and IFSA in Ukraine have been developing Business Incubators, Small Business Support Centres and Credit Unions. IFSA is also conducting a Small Cities Programme.

Programmes in the three countries mentioned collaborate actively with government bodies and authorities at different levels as well as with international aid programmes including NATO, the EU, the Institute for Strategic Studies and OSI NY.

However, Social Adaptation programmes in Poland, Ukraine and Russia are not linked to military property ownership restructuring. Thus revenue completed from the sale of military properties is not directly transferred to officers and their families. It could be an important financial input because other sources are poor.

The necessity of conservation of the base structure as well as the environmental clean-up proceeding conversion may provide job places for officers and civilians who worked at the military base. In the first stage, all enterprises and businessmen wishing to create business and meeting the basic requirements of the 'bible' should be accepted into the military base zone. Latterly, local authorities can meet higher requirements for enterprises, in particular, by giving priority to those which tend to self-development, and hence, create real preconditions for the development of an industrial zone in the military base territory.

There are many reasons why training, consulting, access to financing, is needed for businesses founded by the military:

1. Retired military personnel in Eastern European countries should learn how to adapt to the new reality, the market economy. Their salaries and life quality are not protected fully by the Army or the Government.
2. Business skills can help the military to use their advantages usefully in business: strategy thinking, team building, responsibility, logistic, etc.
3. The recent slowing down of economies makes small enterprises unable to do well, often leading to bankruptcy.
4. Companies established by military men are usually not innovative and thus not competitive. It would be better if they had access to high technology available in the Army and use this potential for the benefit of their business and the Army.
5. For men, the psychological problems arise when their wives are forced to go to work and supplement the family budget. It makes a new position in civilian society which is not easily acceptable.
6. Military men very often tend to relate their companies to military industry and collaborate with other military men in companies. They prefer to stay in the same group of other officers and soldiers and cultivate old customs. It is not a prospect in the free market over a long period of time. External partners seem to be less valuable.

3. SME's needs to cope with the Civilian Market

Having had the opportunity to work on business incubation in countries in transition one is repeatedly struck by the similarity of the problems confronting entrepreneurs in these countries and developed countries, such as:

- Lack of affordable premises with just the amount and type of space required;
- Lack of seed capital and/or working capital;
- Regulatory barriers and constraints; and, generally,
- Lack of support systems, network or appropriate business services.

Perhaps the most important function the Business Incubators and Small Business Support Centre can provide, regardless of its size, is the linkage to the various types of new business needs. While many communities in Western countries already offer different types of assistance, the assistance in Eastern European countries is often poor and not very visible and thus loses a good deal of its effectiveness. Providing a visible linking function will also assist the programme by identifying weak or non-existent services that may be necessary for a strong business incubation programme.

While the programme addresses such needs as management assistance and small business loans have been provided, little emphasis was placed on bringing them together into a cohesive system. Many different operations could be deemed successful despite widely-varying services. The reasonable conclusion derived from these findings is that there must be some important, yet intangible, factor, which contributes to the success of a small business programme.

This may be simply the provision of a suitable business development climate. By bringing needed services together in an environment in which entrepreneurs can interact and assist one another the new business development process seems to acquire synergy. In most of the business and innovation centres where a wide variety of small firms are housed, it appears that just the ability of new entrepreneurs to interact regularly with other businesses enhances the chance of success for all of the tenants.

This need for interaction could also explain why most of business incubator tenants, which outgrow the facility, tend to locate near it when selecting a new location.

The Business Incubators, Small Business Support Centres and Credit Unions conducted by the Open Society Institute in Ukraine and Russia which are underway, aim at solving local economic and social problems by:

- Training entrepreneurs, military and their families in how to start a business activity and develop enterprise in a free market economy,
- Search for and evaluation of new business projects,
- Technical, economical, financial, legal and other forms of consulting,
- Search for finances for business projects and making acquisition of capital easier,
- Business planning,
- Technology transfer,
- Use of idle facilities providing an opening for new businesses.

These programmes are not focused exclusively on the military and families. The services of Business Incubators and Business Support Centres are offered to the whole community. A wide range of beneficiaries help the military to adapt to the civilian community by working and making business together.

In Poland, the Military Centre for Professional Activization has just started similar programmes.

Typically, there are three major areas in which public action can be taken to influence the economic development process:

- Improving the business climate,
- Investing in people,
- Providing infrastructure.

As with any tool, the primary issue concerning Business Centres' effectiveness is whether they help accomplish the economic development strategy.

Once the objectives of the region are selected and the pool of potential new businesses has been surveyed and analyzed within the local economy, the system of delivery of business development services can be established.

While the choice of whether BC is the proper economic development tool to serve local economic development needs is a critical one, equally important is an understanding of what conditions BC is intended to remedy. It is not sufficient to say that the BC purpose is to create new firms and jobs. There must be some reasons why the local economy is not already doing that, given that in some places in the country, new businesses are formed at rates much greater than in other places.

Having determined the economic development role the incubator will play - that is, what local economic conditions it will focus on - you should then consolidate these objectives into an overall statement describing the BC purposes and primary approach in addressing local economic conditions.

In Poland the new opportunities create an offset programme. The Polish Government intends to use it for technology transfer.

4. Conclusions

1. Business Incubators, Small Business Support Centres and Credit Unions/Entrepreneurship Loan Funds are tools used in community economic development. Therefore the Centres should correspond to local development conditions and needs.
2. Economic development is the process of making investments that produce goods and services for local, regional, national and/or international consumption. The intended outcome of economic development is the creation of new and additional wealth for people and places.
3. The public interest in economic development becomes one of remedying negative conditions (unemployment, low per capita income, base closure, etc.) and creating an environment where either public and/or private investment decisions will generate economic opportunities and wealth for all people and places.
4. Governments and armies in Eastern Europe should care much more about retired and released military personnel than before. They can develop effective programmes on rendering assistance to people who lost their jobs in connection with base closure; work out ways of support for local communities; seek for organizations and companies which would like to take responsibility for community economic adjustment.
5. Vocational training for retired military personnel in Eastern European countries is not enough. Eastern Europe needs more active programmes for entrepreneurship promotion.
6. Decision-makers should learn from Western countries how to support retired military personnel. Study tours and joint projects are needed. Study tours to the USA for Ukrainians, Russians and Bulgarians, hosted by Tom Shea, are a good example. The IFSA in Ukraine developed more effective collaboration with the National Coordination Centre for Social and Professional Adaptation of Servicemen Transferred to the Reserves or Retired (NCC).
7. Eastern European countries should collaborate and develop a network to exchange experience and know-how regarding social adaptation and small business development programmes.
8. New NATO countries should adapt NATO standards regarding the policy towards retired and released military personnel as well as military base closure.
9. Companies established by the military have the advantage of easier access to high technology. However it should be allowed to be used in private business.
10. Military businessmen should focus on the whole market rather than exclusively on collaboration with the Army.
11. Military base closure needs different activities which can be accomplished by retired military companies. Facilities should be offered to develop business incubators and industrial zones.
12. Revenues collected from selling military property should be used to support military men in small business: loans, technical assistance, training, etc.
13. The aim of the network of discharged and retired military personnel programmes should be to increase knowledge about solving the problems of retired personnel, business incubation, economic policies and development, entrepreneurship, as well as technology transfer by acquiring mutual projects to:
    - Develop the exchange of information and experience in the fields of regional development, BI, SBSC, CU/ELF
    - Strengthen their co-operation
    - Transfer know-how and information in the field of business incubation, management and small and medium companies

- Develop training programmes for managers of small and medium-sized companies managed by military personnel
- Use offset opportunities

14.The criterion of conversion, the quantity of newly created jobs, requires more precise definition and application in Eastern European countries, as there is a real lack of enterprise on the disbursement of wages and the existing level of the minimum wage does not provide a living for workers. In Eastern Europe, newly-created jobs, providing at least a standard of living to workers, can be accepted as the criterion of successful conversion.

Marketing of Dual Use Technologies and Problems of Distribution of Rights in Intellectual Property in the Defence-Industrial Complex of Russia

A.M. Zherebin
Russia

Abstract. The report highlights two closely interconnected problems that Russian small and medium enterprises (SME's) employing dual use technologies encounter today.

The first problem is connected with the key question of marketing dual use technologies. The second is rights distribution of intellectual property taking into account State interests.

1. Marketing of dual use technologies

In Russia there are all the basic preconditions for a transfer of the economy up the innovation path of development. First of all, there is a sufficiently developed scientific basis, great technological potential and an advanced educational system. Today many scientific and research organizations work actively in pursuing transformation of knowledge into competitive production or increasing enterprise profits in the period of DIC restructuring. World practice has already developed a most effective method of translation of new knowledge into the market-oriented commodities. This method was named technology marketing. It is designed to provide for all the innovation process scientists and developers, workers and investors - to become commercial partners and to gain from the realization of intellectual activity results (IAR) and new science-intensive production. Such a scheme provides every participant in economic circulation with a high motivation for achieving success and a high speed of project realization. Especially important and significant for successful development and transfer of intellectual activity results to industrial use is the taking into account of the interests of small-scale enterprises. SME's do not only have smaller capital in comparison with large companies and cannot support the financing of large advertising campaigns, but also often do not have qualified personnel to ensure business activities both in Russia and in world markets.

Development of a technologies marketing mechanism that can be applied to the contemporary economic situation in Russian requires taking into account four important features characteristic of technological innovation.

1.1 The first feature

The first feature is that not every development becomes a commodity. The world practice of today proves the following: from 100 developments of an applied character, which try to achieve the objective of turning into a competitive commodity, only 6-8 reach the market. At the same time it is known that as many as half of them yield the expected economic effect. Every participant in the economic circulation of dual use technologies, including the State as technology proprietor, must give from the first stage of project development an answer to a question - how much is that very technology economically effective. In cases where the answer is in the affirmative, the venture participants should concentrate the necessary resources and means to ensure a high speed of marketing.

1.2 The second feature

The second feature is the correct structure of resource distribution between different innovation cycle stages. World statistics give the following data as to expenses distribution over all the economic circulation of IAR:

- At a fundamental stage a standard unit of resources is spent;
- To carry out development work and to make a prototype sample, 10 units are necessary;
- To assimilate a development of the sample in manufacture and to bring it to the market, it is necessary to invest 100 units.

And these 100 units are not, as a rule, budget money. It is that very commercial potential that is necessary to draw the form of investments into the innovation projects. Only after having received the answer to a question, from where the money can be drawn for manufacturing and marketing the commodity to the market, is it possible to provide the transformation of scientific research realization into a concrete product which will master certain segments of the market.

1.3 The third feature

The third feature consists in the fact that the rights for intellectual property created during the project realization prove to be the object of marketing. It is a special kind of goods. Its realization requires special legal assistance. The specificity of dual use technologies is the possibility of using them for the creation of techniques for military and special purpose equipment, influencing the defence capacity of the country. In this connection there arises a question of what will be the impact of marketing of technologies and their transfer to the world market on national security and stability in the world. For example, the essential stability dependence of the world on non-proliferation of missile technologies has resulted in prohibition of their proliferation in the world.

1.4 The fourth feature

And, finally, the fourth feature is that marketing of scientific development is in itself a complex technology. For its successful realization we require specially prepared personnel capable of technological management. In Russia there is a very critical problem of how to train specialists in business management for the country's introduction into the world trade system, especially in the field of high technologies. It is known that the project managers are the very people upon whom strongly depends the success of technology marketing, upon whom the investor invests the money, not in the project, but in the people who can realize the project and who can bring commercial success.

The problem of marketing dual use technologies consists is in many respects that the researchers, while trying to attract the attention to their development of manufacturers and investors, do not generally possess an attractive investment index, but operate only by technical parameters like «higher than the world level», «the analogues do not exist» etc. But such declarations serve nothing during talks with commercial partners. A totally different presentation of such offers is necessary for launching a commercial partnership. It is necessary to transform scientific development into scientific goods which can be offered to these partners.

We began work in Russia for the selection of scientific developments having a high degree of marketing. We have developed several Federal target programmes, providing for both scientific development of technologies, and bringing them up to the stage of industrial use. A vivid example of such a successful working programme is «The National Technological Base». Through it, more than 1,000 projects have already passed. What is the final result? About 10 percent of scientific developments can be marketed today. We recommend forming innovation projects on the basis of these developments. For risk reduction for investors we offer to finance such projects on a share basis with the use of budget money and loans from State off-budget funds. As our practice shows, in this case it is possible to draw in significant commercial resources and thereby collect the necessary means for innovation project realization. Budget funds in this case play a role of catalyst, attracting private commercial capital.

Analyzing Russian marketing experience in general, it is possible to note that its basic shortcoming is the absence of a really scientific methodic support of dual use technologies marketing and an effective strategy of involving them in economic circulation and transfer to the world market.

2. Problems of distribution of rights in intellectual property in the defence-industrial complex of Russia

A wide scale practice of involving IAR in economic circulation is necessary to create an enterprise environment within scientific and technological circles. A major element of this environment is small-scale technological firms. A small-scale technological firm in the scientific and technical sphere plays a role of interface between science, industry and market. On one hand, it is bringing scientific developments on the market; on the other hand, it is scanning the market and delivering orders for market-oriented research and scientific developments.

The absence in Russia of effective support of small-scale enterprises is the reason why this sector of the economy is extremely underdeveloped. It also leads to the impossibility of involving dual use technologies created in the DIC in economic circulation.

Further successful development of Russia is more and more closely connected with sustainable innovation development of the economy, involving IAR in economic circulation and incorporating results of scientific, experimental design and technological research. The importance and urgency of research in involving IAR in economic circulation as a result of intellectual activities of scientists and specialists from the scientific and technical sectors under conditions of the developing market economy are explained, first of all, by the fact that only on the basis of use of science achievements, application of effective technologies and innovation manufacturing techniques will the possibility appear of successfully solving sharp economic and social problems existing in Russia. Regretfully we have to note that the share of Russia in world sales of high technology produce is less than 1% of total sales amounting to a sum equal to 2 billion dollars per year.

Dual use technologies are a part of intellectual activities, which have no patent protection ensuring authors' payment for copyrights. Many researchers studying a problem of IAR in general note that the absence of protection and safeguarding of IAR leads to their «plunder and illegal use». At the same time it is necessary to note that absence of legal mechanisms for involving IAR in economic circulation, created with budget money, brings not a smaller, but at times greater harm to the economy and budget of the State because many technical and technological innovations are being sent to storehouses and foundations («on the shelf») instead of being used in innovation manufacturing. In this situation money spent on R&D and technological work does not participate in economic circulation and does not participate in resources returned to the budget.

3. Scientific-methodic support to marketing of dual use technologies

For successful marketing of dual use technologies it is necessary to create its scientific-methodic support.

The strategy of marketing dual use technologies is carried out through licensing, creation of new companies and by fostering various forms of partnership such as social partnership (SP); creation of a technological consortia (TC) or scientific-research consortia (SRC).

An important feature of forming dual use technologies marketing strategy is the necessity of system research. To arrive at this goal it is necessary to develop the following scientific-methodic support base:

- Objectives of strategic marketing planning;
- Methods and principles of strategic planning;
- The strategy of marketing, management, capitalization and marketing of technologies;
- Forming of a number of action alternatives at the stage of technology circulation in conditions of a market competition;
- The choice of optimum strategies of marketing;
- Methods of stimulation and encouragement of technology marketing participants and finding a balance between their interests;
- Monitoring of marketing and prognosis analysis of its development.

As is shown by technology application experience in the US, for effective economic circulation it is necessary to use a system of encouragements, which should provide for fulfilment of the following conditions:

- Balance of interests (of authors, patentees, participants in economic circulation, the State and other participants);
- Predictability and stability of encouragements;
- Taking into account of efforts and risks for economic circulation participants;
- Support by a company management;
- Benevolent climate for enterprise activities;
- Creation of funds for encouragements.

In discussing problems of stimulating dual use technologies marketing one can note that in solving such difficult socio-economic tasks, as in the balancing of interests, mathematical methods of optimisation which find practical application in solving concrete tasks of profits distribution in the oil and gas sectors of the economy have been developed and successfully applied.

In conclusion we should mention some areas of application of a scientific-methodic support for technology marketing. First of all is the formation of a strategy of dual use technologies commercialisation by SME's. Secondly, an assessment of whether the technologies use is in the interests of the State during their transfer to manufacturers or transfer to third persons on a competitive basis. Thirdly, for decision-making, as to the expediency of the protection of the rights in intellectual property on the basis of the received results of intellectual activity. Fourthly, for decision-making, as to the setting up of innovation manufacturing on the basis of use of dual use technologies. Fifthly, for choosing production diversification directions while reforming and developing the DIC on the basis of dual use technologies.

Subject Index

Author Index